New Lipid Lowering Therapies for Cardiovascular and Metabolic Diseases

New Lipid Lowering Therapies for Cardiovascular and Metabolic Diseases: Lessons from the Past and Future Challenges

Special Issue Editor

Ahmed Bakillah

MDPI • Basel • Beijing • Wuhan • Barcelona • Belgrade

Special Issue Editor
Ahmed Bakillah
King Abdullah International Medical
Research Center (KAIMRC)
Saudi Arabia

Editorial Office
MDPI
St. Alban-Anlage 66
4052 Basel, Switzerland

This is a reprint of articles from the Special Issue published online in the open access journal *Diseases* (ISSN 2079-9721) from 2017 to 2019 (available at: https://www.mdpi.com/journal/diseases/special_issues/therapies_CVD)

For citation purposes, cite each article independently as indicated on the article page online and as indicated below:

LastName, A.A.; LastName, B.B.; LastName, C.C. Article Title. *Journal Name* **Year**, *Article Number*, Page Range.

ISBN 978-3-03921-676-5 (Pbk)
ISBN 978-3-03921-677-2 (PDF)

Contents

About the Special Issue Editor

Ahmed Bakillah Research Scientist and Head of the Research Training & Development Section at King Abdullah International Medical Research Center (KAIMRC), Eastern Region- Al Ahsa, Saudi Arabia. Prior to joining KAIMRC, Dr. Bakillah was an Assistant Professor in the Department of Medicine at SUNY Downstate Medical Center, Brooklyn, NY, USA. He also served as Research Director for the Empire Clinical Research Investigator Program (ECRIP). Dr. Bakillah completed his Ph.D. at Rene Descartes University of Paris, France, and performed his postdoctoral fellowship at the Medical College of Pennsylvania, Philadelphia, PA, USA. His research interests lie in the area of cardiovascular diseases with a focus on lipids and lipoprotein abnormalities in diabetes, obesity, and atherosclerosis. Dr. Bakillah worked for almost a decade at two major pharmaceutical companies (Novo Nordisk A/S, Denmark and GlaxoSmithKline, USA). Dr. Bakillah has served for about ten years as the Managing Editor for the journal Nutrition & Metabolism. He currently serves as an ad-hoc reviewer and member of the Editorial Board in several peer-reviewed journals. Dr. Bakillah also served as Guest Editor for special issues of Frontiers in Bioscience (http://www.bioscience.org/special-issue-details?editor_id=1714) and Diseases (https://www.mdpi.com/journal/diseases/special_issues/therapies_CVD). Dr. Bakillah has published numerous scientific papers in highly impacted journals, and his work has been supported by intra- and extramural grants from various agencies in Europe and the USA

Preface to "New Lipid Lowering Therapies for Cardiovascular and Metabolic Diseases: Lessons from the Past and Future Challenges"

This book illustrates some of the most recent research efforts that have been made in lowering plasma cholesterol levels in patients with CVD. Selected articles aimed to illuminate advances and urgent challenges in the management of CVD, including disease management using statin-combined therapeutic strategies.

Ahmed Bakillah
Special Issue Editor

 *diseases*

Review

Anacetrapib, a New CETP Inhibitor: The New Tool for the Management of Dyslipidemias?

Theodosios D. Filippatos *, Anastazia Kei and Moses S. Elisaf

Department of Internal Medicine, School of Medicine, University of Ioannina, Ioannina 45110, Greece; keianastazia@gmail.com (A.K.); melisaf54@gmail.com (M.S.E.)
* Correspondence: filtheo@gmail.com; Tel.: +302651007516

Received: 14 September 2017; Accepted: 29 September 2017; Published: 29 September 2017

Abstract: Cholesteryl ester transfer protein (CETP) inhibitors significantly increase serum high-density lipoprotein cholesterol (HDL) cholesterol levels and decrease low-density lipoprotein cholesterol (LDL) cholesterol concentration. However, three drugs of this class failed to show a decrease of cardiovascular events in high-risk patients. A new CETP inhibitor, anacetrapib, substantially increases HDL cholesterol and apolipoprotein (Apo) AI levels with a profound increase of large HDL2 particles, but also pre-β HDL particles, decreases LDL cholesterol levels mainly due to increased catabolism of LDL particles through LDL receptors, decreases lipoprotein a (Lp(a)) levels owing to a decreased Apo (a) production and, finally, decreases modestly triglyceride (TRG) levels due to increased lipolysis and increased receptor-mediated catabolism of TRG-rich particles. Interestingly, anacetrapib may be associated with a beneficial effect on carbohydrate homeostasis. Furthermore, the Randomized EValuation of the Effects of Anacetrapib Through Lipid-modification (REVEAL) trial showed that anacetrapib administration on top of statin treatment significantly reduces cardiovascular events in patients with atherosclerotic vascular disease without any significant increase of adverse events despite its long half-life. Thus, anacetrapib could be useful for the effective management of dyslipidemias in high-risk patients that do not attain their LDL cholesterol target or are statin intolerable, while its role in patients with increased Lp(a) levels remains to be established.

Keywords: anacetrapib; cholesteryl ester transfer protein; cardiovascular disease; apolipoprotein; diabetes

---

1. Introduction

Cholesteryl ester transfer protein (CETP) is a glycoprotein synthesized mainly in the liver, which plays a prominent role in the bidirectional transfer of cholesterol esters and triglycerides (TRG) between lipoproteins, that is the transfer of cholesterol esters from the cardioprotective high-density lipoprotein (HDL) particles to the potentially atherogenic non-HDL particles (very low-density lipoprotein (VLDL) particles, remnant lipoproteins and low-density lipoprotein (LDL) particles) [1,2]. Drugs that inhibit CETP are able to increase HDL cholesterol and also to decrease serum LDL cholesterol levels [3]. However, the randomized placebo-controlled trials that evaluated the effects of three drugs of this class, that is torcetrapib, dalcetrapib and evacetrapib, failed to show a beneficial cardiovascular effect [4–7]. These trials were prematurely terminated due to either off-target toxicity (torcetrapib) or lack of efficacy (dalcetrapib, evacetrapib) [4–8]. However, it has been recently reported that in the Randomized EValuation of the Effects of Anacetrapib Through Lipid-modification (REVEAL) trial, anacetrapib, a new CETP inhibitor, led to a significant decrease of cardiovascular events.

Thus, the aim of the review is to present the lipid/lipoprotein and cardiovascular effects of anacetrapib, as well as to discuss possible future indications of the drug.

2. Main Results of the REVEAL Trial

The REVEAL trial was a randomized placebo-controlled trial that assessed the efficacy and safety of anacetrapib in 30,449 adults with atherosclerotic vascular disease on intensive atorvastatin treatment (mean LDL cholesterol 61 mg/dL and mean non-HDL cholesterol 92 mg/dL) [9]. The administration of 100 mg of anacetrapib for a mean follow-up period of 4.1 years was associated with a significant reduction of the primary end point (first major coronary event) by 9% (rate ratio 0.91, 95% confidence interval 0.85–0.97, p = 0.004). This risk reduction was evident even though patients treated with anacetrapib exhibited slightly higher systolic blood pressure (SBP) and diastolic blood pressure (DBP) values (by 0.7/0.3 mm Hg, respectively) at the final visit compared with the control group. Treatment was well tolerated, and no significant differences between groups in the risk of death, cancer or other serious adverse events were observed. In fact, the incidence of new-onset diabetes mellitus among patients without diabetes mellitus at baseline was lower in the anacetrapib group as compared to the control group (5.3% vs. 6%, rate ratio 0.89, 95% CI 0.79–1.00, p = 0.0496) [9].

2.1. Effects of Anacetrapib on Lipid and Lipoprotein Profile

Anacetrapib is an inhibitor of CETP, which can lead to impressive changes of the serum lipid profile, as shown in Table 1, which includes the results of the REVEAL trial and two randomized trials that evaluated the efficacy of the drug in both high-risk patients, as well as in patients with familial hypercholesterolemia [9–11].

Table 1. Anacetrapib (100 mg/d)-associated percentage changes of lipid parameters versus placebo in randomized clinical trials.

Lipid Parameters	REALIZE Trial [11] (Patients with Familial Hypercholesterolemia)	DEFINE Trial [10] (Patients with Coronary Heart Disease or at Risk for Coronary Heart Disease)	REVEAL Trial [9] (Patients with Atherosclerotic Vascular Disease) *
LDL cholesterol	−39.7% **	−39.8% ***	−41% (direct method) −17% ** (in a subgroup of 2000 patients)
HDL cholesterol	+102.1%	+138.1%	+104%
Non-HDL cholesterol	−36.4%	−31.7%	−18%
Apo B	−24.8%	−21%	−18%
Apo AI	+32.9%	+44.7%	+36%
Lp(a)	−27.9%	−36.4%	−25%
TRG	−5.5%	−6.8%	−7%

* Lipid levels at the trial midpoint. ** Measured by β quantification. *** LDL-cholesterol levels were calculated with the use of the Friedewald equation: LDL cholesterol = total cholesterol − (HDL cholesterol + [triglycerides ÷ 5]). If the triglyceride level was more than 400 mg per deciliter, LDL-cholesterol was measured by means of preparative ultracentrifugation separation. REALIZE: Randomized Evaluation of Anacetrapib Lipid-Modifying Therapy in Patients with Heterozygous Familial Hypercholesterolemia, DEFINE: Determining the EFficacy and Tolerability of CETP INhibition with AnacEtrapib, REVEAL: Randomized EValuation of the Effects of Anacetrapib Through Lipid-modification, LDL: low-density lipoprotein, HDL: high-density lipoprotein, Apo: apolipoprotein, Lp(a): lipoprotein (a), TRG: triglycerides.

Beyond the prominent increase of HDL cholesterol, a marked decrease of LDL cholesterol was observed in these trials. Even though the underlying mechanisms of this increase in LDL cholesterol are not clear, kinetic data of lipid metabolism during anacetrapib administration in humans points to the following potential mechanisms (Figure 1) [12–15]:

(1) Increased catabolism of LDL particles due to anacetrapib-associated compositional changes of LDL particles, such as increased TRG content and particle size, leading to increased LDL particles' affinity to the LDL receptors [15].

(2) Decreased proprotein convertase subtilisin/kexin type 9 (PCSK9) levels (a CETP-independent mechanism) leading to decreased LDL receptors' degradation and, therefore, increased LDL receptors' activity and LDL particles' catabolism [14,16].

(3) Decreased drug-associated cellular cholesterol concentration in the liver, due to increased cholesterol efflux, which results in the activation of sterol regulatory element-binding protein 2 (SREBP-2) leading to increased synthesis of LDL receptors [15].

(4) Reduced expression of the inducible degrader of the LDL receptors (IDOL) due to decreased hepatic oxysterols resulting in a reduced activation of liver X receptors (LRX). This reduced IDOL expression is associated with increased LDL receptors' activity and subsequently with increased catabolism of LDL particles [15].

(5) Reduced transfer of cholesterol esters from HDL to LDL is also associated with a reduction of LDL cholesterol, though this mechanism cannot explain the observed marked reduction of apolipoprotein (Apo) B levels that reflects a reduction in the concentration of LDL particles [12,15].

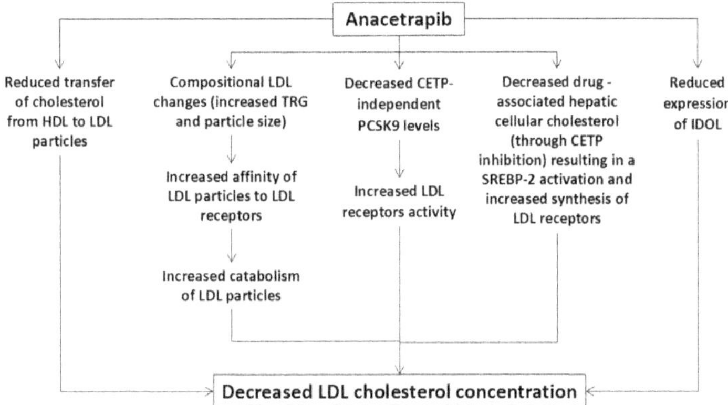

Figure 1. Potential mechanisms of anacetrapib-mediated reduction of low-density lipoprotein (LDL) cholesterol. HDL: high-density lipoprotein, LDL: low-density lipoprotein, CETP: cholesteryl ester transfer protein, TRG: triglycerides, SREBP: sterol regulatory element binding protein, IDOL: inducible degrader of the low-density lipoprotein receptor, PCSK9: proprotein convertase subtilisin/kexin type 9.

Thus, many mechanisms may explain the LDL cholesterol-lowering efficacy of anacetrapib. LDL cholesterol is an established cardiovascular risk factor and represents the main target of hypolipidemic therapy [17,18]. Furthermore, many of the above mechanisms explain the anacetrapib-associated lowering of Apo B levels (Table 1). In this context, the findings of the REVEAL trial were in accordance with the results of the Cholesterol Treatment Trialists Collaboration meta-analysis (n = 170,000 participants) [19], which showed that a reduction of non-HDL cholesterol (which represents the Apo B-containing particles) by 17 mg/dL is expected to reduce the rate of major coronary events by 10%. In fact, recent Mendelian randomization studies have shown that Apo B is a better than LDL cholesterol predictor of an increased cardiovascular risk in patients with gene variants that reflect combined CETP inhibitor and statin treatment [20]. Therefore, it has been proposed that the anacetrapib-mediated decrease of Apo B is the main mechanism of cardiovascular risk reduction in patients taking a statin [20].

A decrease of lipoprotein a (Lp(a)) levels was repeatedly observed after anacetrapib and other CETP inhibitors' administration (Table 1) [9–11]. Interestingly, a recently published study clearly showed that the anacetrapib-mediated reduction of Lp(a) levels (by 34.1%) is due to a reduction of the Apo (a) production rate (by 41%) and not due to changes of Apo (a) fractional catabolic rate [21]. Lp(a) is an established cardiovascular risk factor [22], thus, the anacetrapib-mediated decrease of its serum concentration may have played a significant role in the positive results of the REVEAL trial.

Additionally, a small decrease of serum TRG levels is also observed with anacetrapib [9–11] mainly due to increased catabolism of TRG-rich VLDL particles (Figure 2). The underlying mechanisms include:

(1) Increased lipolysis of the large TRG-rich VLDL particles through lipoprotein lipase (even without increased lipolytic activity) [23].

(2) Other compositional changes in VLDL particles, such as increased Apo E and reduced Apo CIII content, which can increase lipoprotein lipase activity and the hepatic receptor-mediated clearance of remnant particles [12,23].

(3) Increased hepatic uptake of the large triglyceride-rich (depleted of cholesterol) VLDL particles [16,23].

(4) CETP-independent decreased PCSK9 levels leading to increased LDL receptors' activity and catabolism of VLDL particles and their remnants [16].

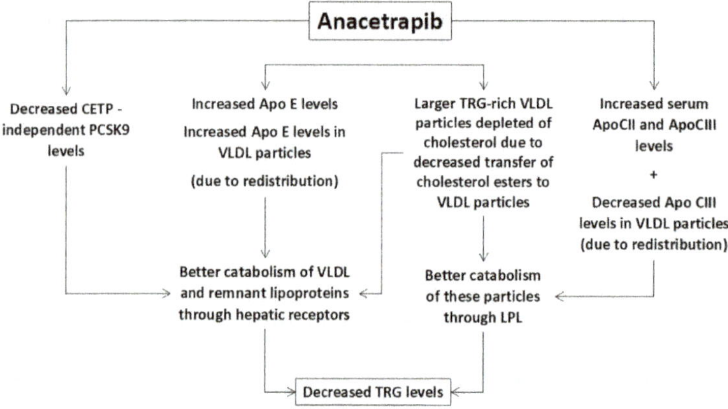

Figure 2. Potential mechanisms of the anacetrapib-mediated reduction of triglycerides (TRG). PCSK9: proprotein convertase subtilisin/kexin type 9, CETP: cholesteryl ester transfer protein, Apo: apolipoprotein, LPL: lipoprotein lipase, VLDL: very low-density lipoprotein.

Finally, a number of trials has delineated the anacetrapib-mediated changes of HDL particles, which include a marked increase of HDL cholesterol, an increase of Apo AI levels (due to its decreased catabolism) and a lesser increase of Apo AII levels, as well as an increase of large cholesterol-rich alpha 2 HDL particles (HDL2 particles) and pre-beta HDL particles, which can lead to an increased ATP-binding cassette transporter (ABCA1)-mediated cholesterol efflux (Figure 3) [24].

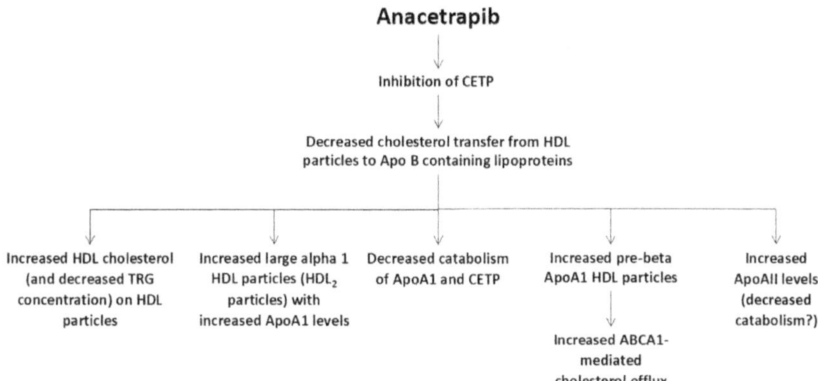

Figure 3. Potential mechanisms of anacetrapib-mediated changes of high-density lipoprotein (HDL) metabolism. CETP: cholesteryl ester transfer protein, Apo: apolipoprotein, TRG: triglycerides, ABCA1: ATP-binding cassette transporter ABCA1 (member 1 of human transporter sub-family ABCA).

2.2. Other Effects of Anacetrapib

In contrast with the other drugs of this class, anacetrapib does not seem to exhibit significant "on-target" adverse effects, such as dysfunctional HDL particles or changes in apolipoproteins that promote atherogenesis, or "off target" effects, such as a significant increase of blood pressure or C-reactive protein levels (which is an inflammatory marker) [8,25]; these adverse effects have been implicated in the negative or neutral cardiovascular effects of other CETP inhibitors. However, in the REVEAL trial, slightly higher SBP and DBP values (by 0.7/0.3 mm Hg, respectively) were observed at the final visit in anacetrapib-treated patients compared with the control group. Alternatively, any on-target or off-target detrimental effects of anacetrapib may have been counterbalanced by the marked decrease of Apo B-containing atherogenic lipoproteins, such as LDL, but also Lp(a) [9].

Mendelian randomization trials have shown that hypolipidemic drugs that reduce LDL receptors' activity are associated with a detrimental effect on carbohydrate homeostasis [26–29], which has received much attention taking into account the wide use of these hypolipidemic drugs. The beneficial effect of CETP inhibitors on glucose homeostasis, as shown in the REVEAL trial and previous studies [9,30], can possibly counteract the dysglycemic effect of other hypolipidemic drugs. Thus, anacetrapib could be a promising therapeutic strategy as an add-on therapy to current hypolipidemic treatment especially in high-risk patients with disorders of carbohydrate metabolism.

It should be mentioned that anacetrapib has a long half-life, which is associated with advantages concerning the patients' compliance to therapy, but may be a disadvantage concerning side effects that can be observed after long-term treatment [31].

3. Possible Future Indications of Anacetrapib

Based on the current knowledge regarding its effects on lipidemic profile and cardiovascular risk, anacetrapib could be useful for the treatment of:

(1) High-risk patients who do not attain their LDL cholesterol target despite optimal hypolipidemic treatment. Currently, these patients could take a PCSK9 inhibitor [32,33], but anacetrapib could be alternatively used taking into account that it is associated with a reduction of cardiovascular events, is given orally once per day (in contrast with the parenteral administration of current PCSK9 inhibitors) and probably will have a much lower cost.

(2) Patients with familial hypercholesterolemia with difficulty achieving their LDL cholesterol target despite optimal hypolipidemic treatment, in whom anacetrapib is associated with a significant improvement of lipid and lipoprotein profile (Table 1).

(3) Patients with statin intolerance, in whom the use of the expensive and parenteral PCSK9 inhibitors is not easily accepted. This possible indication of anacetrapib could be the most important in the future since non-statin treatment with a potent CETP inhibitor (possibly in combination with ezetimibe) could be associated with large concordant absolute reductions of both LDL cholesterol and Apo B and, therefore, of cardiovascular risk [20].

(4) Patients with increased serum Lp(a) levels, although more data are needed in this population.

It is worth mentioning that two new CETP inhibitors that induce significant beneficial changes in lipidemic profile (including a reduction of Apo B-containing lipoproteins) are under investigation [Obicetrapib (AMG-899, TA-8995) and K-312] [34,35]. The results of the relative studies will show whether there is a class effect of CETP inhibitors on cardiovascular events and will help to delineate the implicated mechanisms.

4. Conclusions

Anacetrapib is associated with significant improvement of lipid and lipoprotein profile and a significant reduction of cardiovascular events in high-risk patients on optimal statin treatment. These data may suggest that anacetrapib could be a useful therapeutic option for the effective hypolipidemic treatment of high-risk patients especially if the results of ongoing trials with other CETP inhibitors confirm the positive cardiovascular effects of this drug class.

Author Contributions: T.F. conceived and wrote the manuscript, A.K. searched the bibliography and made critical revisions and M.E. conceived the manuscript and made critical revisions.

Conflicts of Interest: This review was conducted independently; no funding was received for writing this manuscript or for covering the costs to publish in open access. M.E. reports personal fees from Astra-Zeneca, grants and personal fees from Merck Sharp & Dohme, personal fees from Pfizer, Abbott, Sanofi-Aventis, Boehringer Ingelheim, Eli-Lilly and GlaxoSmithKline. A.K. and T.F. have given talks and attended conferences sponsored by various pharmaceutical companies, including Bristol-Myers Squibb, Pfizer, Lilly, Abbott, Amgen, AstraZeneca, Novartis, Vianex, Teva and Merck Sharp & Dohme.

References

1. Barter, P.J.; Rye, K.A. Cholesteryl ester transfer protein inhibition as a strategy to reduce cardiovascular risk. *J. Lipid. Res.* **2012**, *53*, 1755–1766. [CrossRef] [PubMed]
2. Tall, A.R. Plasma cholesteryl ester transfer protein. *J. Lipid. Res.* **1993**, *34*, 1255–1274. [PubMed]
3. Barter, P.J.; Rye, K.A. Cholesteryl Ester Transfer Protein Inhibition Is Not Yet Dead-Pro. *Arterioscler Thromb Vasc. Biol.* **2016**, *36*, 439–441. [CrossRef] [PubMed]
4. Barter, P.J.; Caulfield, M.; Eriksson, M.; Grundy, S.M.; Kastelein, J.J.; Komajda, M.; Lopez-Sendon, J.; Mosca, L.; Tardif, J.C.; Waters, D.D.; et al. Effects of torcetrapib in patients at high risk for coronary events. *N. Engl. J. Med.* **2007**, *357*, 2109–2122. [CrossRef] [PubMed]
5. Schwartz, G.G.; Olsson, A.G.; Abt, M.; Ballantyne, C.M.; Barter, P.J.; Brumm, J.; Chaitman, B.R.; Holme, I.M.; Kallend, D.; Leiter, L.A.; et al. Effects of dalcetrapib in patients with a recent acute coronary syndrome. *N. Engl. J. Med.* **2012**, *367*, 2089–2099. [CrossRef] [PubMed]
6. Lincoff, A.M.; Nicholls, S.J.; Riesmeyer, J.S.; Barter, P.J.; Brewer, H.B.; Fox, K.A.A.; Gibson, C.M.; Granger, C.; Menon, V.; Montalescot, G.; et al. Evacetrapib and Cardiovascular Outcomes in High-Risk Vascular Disease. *N. Engl. J. Med.* **2017**, *376*, 1933–1942. [CrossRef] [PubMed]
7. Filippatos, T.D.; Elisaf, M.S. Evacetrapib and cardiovascular outcomes: Reasons for lack of efficacy. *J. Thoracic Disease* **2017**, *9*, 2308–2310. [CrossRef] [PubMed]
8. Filippatos, T.D.; Klouras, E.; Barkas, F.; Elisaf, M. Cholesteryl ester transfer protein inhibitors: Challenges and perspectives. *Expert. Rev. Cardiovasc. Ther.* **2016**, *14*, 953–962. [CrossRef] [PubMed]
9. HPS3/TIMI55-REVEAL Collaborative Group. Effects of Anacetrapib in Patients with Atherosclerotic Vascular Disease. *N. Engl. J. Med.* **2017**, *377*, 1217–1227.

10. Cannon, C.P.; Shah, S.; Dansky, H.M.; Davidson, M.; Brinton, E.A.; Gotto, A.M.; Stepanavage, M.; Liu, S.X.; Gibbons, P.; Ashraf, T.B.; et al. Safety of anacetrapib in patients with or at high risk for coronary heart disease. *N. Engl. J. Med.* **2010**, *363*, 2406–2415. [CrossRef] [PubMed]
11. Kastelein, J.J.; Besseling, J.; Shah, S.; Bergeron, J.; Langslet, G.; Hovingh, G.K.; Al-Saady, N.; Koeijvoets, M.; Hunter, J.; Johnson-Levonas, A.O.; et al. Anacetrapib as lipid-modifying therapy in patients with heterozygous familial hypercholesterolaemia (REALIZE): A randomised, double-blind, placebo-controlled, phase 3 study. *Lancet* **2015**, *385*, 2153–2161. [CrossRef]
12. Millar, J.S.; Lassman, M.E.; Thomas, T.; Ramakrishnan, R.; Jumes, P.; Dunbar, R.L.; de Goma, E.M.; Baer, A.L.; Karmally, W.; Donovan, D.S.; et al. Effects of CETP inhibition with anacetrapib on metabolism of VLDL-TG and plasma apolipoproteins C-II, C-III, and E. *J. Lipid Res.* **2017**, *58*, 1214–1220. [CrossRef] [PubMed]
13. Krauss, R.M.; Pinto, C.A.; Liu, Y.; Johnson-Levonas, A.O.; Dansky, H.M. Changes in LDL particle concentrations after treatment with the cholesteryl ester transfer protein inhibitor anacetrapib alone or in combination with atorvastatin. *J. Clin. Lipidol.* **2015**, *9*, 93–102. [CrossRef] [PubMed]
14. Barter, P.J.; Tabet, F.; Rye, K.A. Reduction in PCSK9 levels induced by anacetrapib: An off-target effect? *J. Lipid Res.* **2015**, *56*, 2045–2047. [CrossRef] [PubMed]
15. Millar, J.S.; Reyes-Soffer, G.; Jumes, P.; Dunbar, R.L.; deGoma, E.M.; Baer, A.L.; Karmally, W.; Donovan, D.S.; Rafeek, H.; Pollan, L.; et al. Anacetrapib lowers LDL by increasing ApoB clearance in mildly hypercholesterolemic subjects. *J. Clin. Investig.* **2015**, *125*, 2510–2522. [CrossRef] [PubMed]
16. Van der Tuin, S.J.; Kuhnast, S.; Berbee, J.F.; Verschuren, L.; Pieterman, E.J.; Havekes, L.M.; van der Hoorn, J.W.; Rensen, P.C.; Jukema, J.W.; Princen, H.M.; et al. Anacetrapib reduces (V)LDL cholesterol by inhibition of CETP activity and reduction of plasma PCSK9. *J. Lipid Res.* **2015**, *56*, 2085–2093. [CrossRef] [PubMed]
17. Filippatos, T.D.; Elisaf, M.S. Are lower levels of LDL-cholesterol really better? Looking at the results of IMPROVE-IT: Opinions of three experts—III. *Hellenic. J. Cardiol.* **2015**, *56*, 7–9. [PubMed]
18. Catapano, A.L.; Graham, I.; De Backer, G.; Wiklund, O.; Chapman, M.J.; Drexel, H.; Hoes, A.W.; Jennings, C.S.; Landmesser, U.; Pedersen, T.R.; et al. 2016 ESC/EAS Guidelines for the Management of Dyslipidaemias. *Eur. Heart J.* **2016**, *37*, 2999–3058. [CrossRef] [PubMed]
19. Baigent, C.; Blackwell, L.; Emberson, J.; Holland, L.E.; Reith, C.; Bhala, N.; Peto, R.; Barnes, E.H.; Keech, A.; Simes, J.; et al. Efficacy and safety of more intensive lowering of LDL cholesterol: A meta-analysis of data from 170,000 participants in 26 randomised trials. *Lancet* **2010**, *376*, 1670–1681. [PubMed]
20. Ference, B.A.; Kastelein, J.J.P.; Ginsberg, H.N.; Chapman, M.J.; Nicholls, S.J.; Ray, K.K.; Packard, C.J.; Laufs, U.; Brook, R.D.; Oliver-Williams, C.; et al. Association of Genetic Variants Related to CETP Inhibitors and Statins With Lipoprotein Levels and Cardiovascular Risk. *JAMA* **2017**, *318*, 947–956. [CrossRef] [PubMed]
21. Thomas, T.; Zhou, H.; Karmally, W.; Ramakrishnan, R.; Holleran, S.; Liu, Y.; Jumes, P.; Wagner, J.A.; Hubbard, B.; Previs, S.F.; et al. CETP (Cholesteryl Ester Transfer Protein) Inhibition With Anacetrapib Decreases Production of Lipoprotein(a) in Mildly Hypercholesterolemic Subjects. *Arterioscler Thromb. Vasc. Biol.* **2017**, *37*, 1770–1775. [CrossRef] [PubMed]
22. Milionis, H.J.; Filippatos, T.D.; Loukas, T.; Bairaktari, E.T.; Tselepis, A.D.; Elisaf, M.S. Serum lipoprotein(a) levels and apolipoprotein(a) isoform size and risk for first-ever acute ischaemic nonembolic stroke in elderly individuals. *Atherosclerosis* **2006**, *187*, 170–176. [CrossRef] [PubMed]
23. Brown, A.L.; Brown, J.M. Anacetrapib-driven triglyceride lowering explained: The fortuitous role of CETP in the intravascular catabolism of triglyceride-rich lipoproteins. *J. Lipid Res.* **2017**, *58*, 1031–1032. [CrossRef] [PubMed]
24. Reyes-Soffer, G.; Millar, J.S.; Ngai, C.; Jumes, P.; Coromilas, E.; Asztalos, B.; Johnson-Levonas, A.O.; Wagner, J.A.; Donovan, D.S.; Karmally, W.; et al. Cholesteryl Ester Transfer Protein Inhibition With Anacetrapib Decreases Fractional Clearance Rates of High-Density Lipoprotein Apolipoprotein A-I and Plasma Cholesteryl Ester Transfer Protein. *Arterioscler Thromb. Vasc. Biol.* **2016**, *36*, 994–1002. [CrossRef] [PubMed]
25. Raal, F.J.; Blom, D.J. Anacetrapib in familial hypercholesterolaemia: Pros and cons. *Lancet* **2015**, *385*, 2124–2126. [CrossRef]
26. Filippatos, T.D.; Filippas-Ntekouan, S.; Pappa, E.; Panagiotopoulou, T.; Tsimihodimos, V.; Elisaf, M.S. PCSK9 and carbohydrate metabolism: A double-edged sword. *World J. Diabetes* **2017**, *8*, 311–316. [CrossRef] [PubMed]

27. Lotta, L.A.; Sharp, S.J.; Burgess, S.; Perry, J.R.B.; Stewart, I.D.; Willems, S.M.; Luan, J.; Ardanaz, E.; Arriola, L.; Balkau, B.; et al. Association Between Low-Density Lipoprotein Cholesterol-Lowering Genetic Variants and Risk of Type 2 Diabetes: A Meta-analysis. *JAMA* **2016**, *316*, 1383–1391. [CrossRef] [PubMed]

28. Filippatos, T.D.; Elisaf, M.S. Effects of ezetimibe/simvastatin combination on metabolic parameters. *Int. J. Cardiol.* **2016**, *202*, 273–274. [CrossRef] [PubMed]

29. Filippatos, T.D.; Elisaf, M.S. Pitavastatin and carbohydrate metabolism: What is the evidence? *Expert Rev. Clin. Pharmacol* **2016**, *9*, 955–960. [CrossRef] [PubMed]

30. Barter, P.J.; Rye, K.A.; Tardif, J.C.; Waters, D.D.; Boekholdt, S.M.; Breazna, A.; Kastelein, J.J. Effect of torcetrapib on glucose, insulin, and hemoglobin A1c in subjects in the Investigation of Lipid Level Management to Understand its Impact in Atherosclerotic Events (ILLUMINATE) trial. *Circulation* **2011**, *124*, 555–562. [CrossRef] [PubMed]

31. Borghi, C.; Cicero, A.F. Pharmacokinetic drug evaluation of anacetrapib for the treatment of dyslipidemia. *Expert Opin. Drug Metab. Toxicol.* **2017**, *13*, 205–209. [CrossRef] [PubMed]

32. Sabatine, M.S.; Giugliano, R.P.; Keech, A.C.; Honarpour, N.; Wiviott, S.D.; Murphy, S.A.; Kuder, J.F.; Wang, H.; Liu, T.; Wasserman, S.M.; et al. Evolocumab and Clinical Outcomes in Patients with Cardiovascular Disease. *N. Engl. J. Med.* **2017**, *376*, 1713–1722. [CrossRef] [PubMed]

33. Filippatos, T.D.; Kei, A.; Rizos, C.V.; Elisaf, M.S. Effects of PCSK9 Inhibitors on Other than Low-Density Lipoprotein Cholesterol Lipid Variables. *J. Cardiovasc. Pharmacol Ther.* **2017**. [CrossRef] [PubMed]

34. Hovingh, G.K.; Kastelein, J.J.; Van Deventer, S.J.; Round, P.; Ford, J.; Saleheen, D.; Rader, D.J.; Brewer, H.B.; Barter, P.J. Cholesterol ester transfer protein inhibition by TA-8995 in patients with mild dyslipidaemia (TULIP): A randomised, double-blind, placebo-controlled phase 2 trial. *Lancet* **2015**, *386*, 452–460. [CrossRef]

35. Miyosawa, K.; Watanabe, Y.; Murakami, K.; Murakami, T.; Shibata, H.; Iwashita, M.; Yamazaki, H.; Yamazaki, K.; Ohgiya, T.; Shibuya, K.; et al. New CETP inhibitor K-312 reduces PCSK9 expression: A potential effect on LDL cholesterol metabolism. *Am. J. Physiol. Endocrinol. Metab.* **2015**, *309*, E177–190. [CrossRef] [PubMed]

Article

Rosuvastatin Improves Vaspin Serum Levels in Obese Patients with Acute Coronary Syndrome

Hayder M. Al-kuraishy *, Ali I. Al-Gareeb and Ali K. Al-Buhadilly

Department of Pharmacology, Toxicology, and Medicine, College of Medicine Al-Mustansiriyah University, P.O. Box 14132 Baghdad, Iraq; dr.alialgareeb78@yahoo.com (A.I.A.-G.); alikadm1977@yahoo.com (A.K.A.-B.)
* Correspondence: Hayderm36@yahoo.com

Received: 5 December 2017; Accepted: 8 January 2018; Published: 16 January 2018

Abstract: Adipose tissue-derived serine protease inhibitor (vaspin), which has endocrine and local roles in atherosclerosis growth, is also synthesized by adipose tissue; it was found that vaspin was negatively correlated with blood pressure in obese patients, while vaspin levels were decreased in endothelial dysfunction. The aim of the present study was to determine rosuvastatin modulation effects on serum vaspin levels in acute coronary syndrome (ACS) with class I obesity. A total number of seventy patients with acute coronary syndrome previously and currently treated with rosuvastatin was compared to 40 patients with IHD not treated by rosuvastatin as a control. Vaspin serum levels were higher in rosuvastatin-treated patients with acute coronary syndrome compared to the patients with acute coronary syndrome not treated by rosuvastatin, $p < 0.01$. Additionally, in the rosuvastatin-treated group, patients with STEMI showed higher vaspin serum levels compared to NSTEMI $p < 0.01$. Conclusion: Rosuvastatin significantly increases vaspin serum levels in acute coronary syndrome.

Keywords: vaspin; acute coronary syndrome; NSTEMI; STEMI

---

1. Introduction

Acute coronary syndrome (ACS) is a cluster of pathological conditions due to a reduction in coronary blood flow caused by coronary thrombosis and/or atherosclerosis, leading to myocardial ischemia and necrosis [1]. Acute coronary syndrome includes ST-elevation myocardial infarction (STEMI) 30%, non-ST elevation myocardial infarction (NSTEMI) 25%, and unstable angina 38%; these are classified according to electrocardiographic changes in ST-segment [2]. Acute coronary syndrome should be differentiated from stable angina (crescendo angina); in addition, new onset angina should be regarded as part of ACS [3]. Acute myocardial infarction (MI) is known as myocardial cell death (necrosis) because of prolonged myocardial ischemia, STEMI occurs when the coronary artery thrombus is initiated rapidly at coronary vascular wall injury, which can be triggered by many factors, including hypertension, cigarette smoking and dyslipidemia [4]. Unstable angina/non-ST-elevation MI patients usually have numerous vulnerable plaques at risk of disruption and rupture, so platelet aggregation in acute coronary syndromes (ACS) leads to unstable plaque [5]. The damage to the full thickness of the heart muscle is an indicator of increased damage percentage, while the partial thickness of the heart muscle damage is called NSTEMI [6].

ACS is associated with inflammatory and non-inflammatory risk factors; high blood levels of C-reactive protein (CRP) may be linked with risk of coronary artery disease and having a heart attack [7]. In healthy people without hyperlipidemia but with increased CRP levels, the statin drugs rosuvastatin significantly reduced the acute cardiovascular events [8,9].

Visceral Adipose Tissue-Derived Serpin (vaspin) is a novel adipokine that is expressed mainly in visceral white adipose tissue [10]. The up-regulation of vaspin synthesis may signify a response to the antagonizing action of fat-derived proteases that antagonize the insulin action; therefore, up-regulation of vaspin expression may be regarded as a defense mechanism against insulin resistance [11]. Additionally, vaspin was negatively correlated with blood pressure in obese patients, while vaspin levels were decreased in patients with impaired endothelial function [12]. Inflammation was thought to be one of the major causes of early atherosclerosis and its complications [13]. Recently, a study predicted that adipokines including vaspin have a local role in preventing the progression of atherosclerotic growth of [14].

Class I obesity is defined by the WHO as moderate obesity where body mass index (BMI) range is 30–35 kg/m^2; obesity is positively correlated with the incidence of the acute coronary syndrome and vaspin serum levels [15].

Statins are the most broadly used lipid-lowering agents in patients with dyslipidemia, they reduce the cardio-metabolic risk factors, morbidity and mortality of cardiovascular diseases independent of their lipid-lowering effect. These effects were observed along with an augmentation in vaspin serum levels [16].

Rosuvastatin is a HMG-CoA reductase inhibitor indicated for dyslipidemia and approved in 2010 by the FDA for primary prevention of ACS due to a reduction in the cardiovascular risk factors regardless of lipid profile levels [17].

Therefore, the aim of the present study was to determine the effect of rosuvastatin on vaspin serum levels in obese patients with acute coronary syndrome.

2. Patient and Methods

In this cohort study, a total number of seventy patients (50 males, 20 females) previously and currently treated with rosuvastatin—25 patients with unstable angina pectoris (12 male and 13 female), 25 patients with STEMI (22 male and 3 female) and 20 patients with NSTEMI (16 male and 4 female)—were enrolled in the study and compared to 40 patients with IHD not treated by rosuvastatin, as a control. Each patient was clinically examined by the consultant, and the diagnosis was achieved by electrocardiograph ECG, cardiac enzymes, and cardiac Troponin (cTnI). The inclusion criteria were: unstable angina pectoris, STEMI and NSTEMI patients with recent acute myocardial infarction admitted to the coronary care unit (CCU). The exclusion criteria were patients with valvular heart diseases, malignant diseases, acute infection, inflammatory disorders, blood disorders, advanced renal disease, liver disease, smoking and diabetes mellitus. All enrolled patients and controlled subjects gave written informed approval before their participation. The procedures were prepared according to the Declaration of Helsinki. This study was approved by the Clinical Research, Ethical Committee, College of Medicine, Al-Mustansiriyiah University, Baghdad-Iraq.

3. Sample Collections and Anthropometric Profiles

After the interview, medical history, current drug pharmacotherapy and anthropometric measures were examined. Body mass index was estimated as kg/m^2, waist and hip circumferences were recorded, and the waist-hip ratio was calculated. Waist-hip ratio was determined by dividing the waist (cm) by the hip (cm), using a cutoff level <0.85 in females and <0.9 in males [18]. Ten milliliters of venous blood was withdrawn at 9 a.m. after overnight fasting, into a plain tube 5 mL (for routine investigations) and into an EDTA tube 5 mL for vaspin and (cTnI) estimations.

Vaspin and cTnI serum levels were determined by ELISA KIT method (vaspin inhibitor) in pg/mL at 450 nm and human troponin-I (TNNI2) in pg/mL at 450 nm, respectively.

Assessment of lipid profile. triglyceride (TG), total cholesterol (TC) and high-density lipoprotein (HDL) were assessed by specific ELISA kits; from this profile, we can measure the following: [19]

Atherogenic index (AI) = log (TG/HDL), when TG and HDL measured in mmol/L.

LDL = (TC)-(HDL)-(TG)/5.

VLDL = TG/5.

Cardiac risk ratio (CRR) = TC/HDL.

4. Statistical Analysis

Data analysis was done using SPSS (IBM SPSS, Stastics for Window, Version 22.00; 2014, Armonk, NY, USA: IBM Corp.). Results are expressed as mean $\pm$ SD, number and percentage. Unpaired student *t* test was used for estimation of differences between two different groups and one way ANOVA test for estimation the differences among treated groups in terms of 95% confidence interval and *t* value. Pearson correlation coefficient was used to evaluate the correlation of vaspin serum levels with other study parameters. The results were regarded as significant when $p < 0.05$.

5. Results

Baseline characteristics of the present study between the study group (the rosuvastatin-treated group) and the control group (the rosuvastatin-free group) demonstrated non-significant differences in most of the patient variables $p > 0.05$ but, there were significant differences in the presentation of hypertension and current statin therapy $p < 0.0001$, in addition to minor differences regarding other pharmacotherapy $p < 0.05$, Table 1.

Table 1. Baseline characteristics of the present study.

Variables	Rosuvastatin Group	Control Group	*p* Value
	(*n* = 70)	(*n* = 40)	
Age	47.87 $\pm$ 12.72	49.85 $\pm$ 11.83	0.52
Gender			
M:F ratio	(71.42:28.57)%	(22:18) %	-
BMI	31.44 $\pm$ 7.81	32.39 $\pm$ 6.77	0.59
W-H ratio			
Men	1.32 $\pm$ 0.67	1.39 $\pm$ 0.77	0.71
Women	0.88 $\pm$ 0.31	0.91 $\pm$ 0.11	0.67
Hypertension	70 (100%)	20 (100%)	<0.0001 **
ACS			
STEMI	25 (35.71%)	8 (40%)	0.72
NSTEMI	25 (35.71%)	10 (50%)	0.24
UA	20 (28.57%)	2 (10%)	0.08
Troponin positive	67 (95.71%)	18 (90%)	0.32
Troponin negative	3 (4.28%)	2 (10%)	0.32
Dyslipidemia	64 (44.8%)	18 (90%)	0.84
Duration of IHD (years)	8.42 $\pm$ 2.28	7.44 $\pm$ 2.39	0.11
Diabetes mellitus	3 (4.28%)	2 (10%)	0.32
Pharmacotherapy			
Anticoagulant	44 (62.58%)	17 (85%)	0.06
Antiplatelet	70 (100%)	18 (90%)	0.0072
ACEIs	44 (62.58%)	18 (90%)	0.02 *
Statins	70 (100%)	-	<0.0001 **
CCB	22 (31.42%)	12 (60%)	0.02 *
β-blockers	6 (8.57%)	5 (25%)	0.04 *
Insulin	3 (4.28%)	2 (10%)	0.32
Complications	10 (15.71%)	4 (20%)	0.53
Shock	2 (2.85%)	1 (5%)	0.63
Heart failure	5 (3.5%)	2 (10%)	0.67
Cardiac aneurysm	1 (1.42%)	1 (5%)	0.33
Death	2 (2.85%)	1 (5%)	0.33

Data presented as mean $\pm$ SD, number and %, * $p < 0.05$, ** $p < 0.01$ versus control. W-H ratio: waist-hip ratio.

Vaspin serum levels were higher in the study group compared to the control group $p < 0.01$, whereas there were non-significant differences in serum cardiac troponin-I serum level, diastolic blood pressure and blood glucose $p > 0.05$. Moreover, the rosuvastatin-treated group revealed significant differences among other biochemical variables compared to the control group (the rosuvastatin-free group), particularly on high-sensitivity CRP, Table 2.

Table 2. Comparison between study and control groups in vaspin serum levels and other cardio-metabolic risk variables.

Cardio-Metabolic Variables	Study Group (*n* = 70)	Control Group (*n* = 40)	*t* Value	95% CI Upper-Lower Limits	*p* Value
Serum cTn-I pg/mL)	74.54 ± 13.32	77.64 ± 12.22	−0.98	3.33–9.53	0.33
Serum vaspin(pg/mL)	603.83 ± 18.13	542.75 ± 38.95	6.8	79.72–42.33	<0.0001 **
TC (mg/dL)	199.28 ± 20.49	266.43 ± 16.59	−15.1	−134.29	<0.0001 **
TG (mg/dL)	166.83 ± 17.34	254.73 ± 22.82	−15.95	−175.78	<0.0001 **
LDL (mg/dL)	118.80 ± 8.75	164.71 ± 13.59	−14.28	−81.81	<0.0001 **
HDL(mg/dL)	47.11 ± 9.86	50.76 ± 7.34	−1.8	0.43–7.73	0.078
VLDL (mg/dL)	33.36 ± 5.83	50.94 ± 6.42	−11.01	−35.15	<0.0001 **
AI	0.189 ± 0.012	0.341 ± 0.021	−30.95	−0.3	<0.0001 **
CRR	4.23 ± 1.44	5.24 ± 1.98	−2.12	−2.01	0.04 *
SBP (mmHg)	166.54 ± 21.54	155.87 ± 19.63	2.09	21.02–0.31	0.043 *
DBP (mmHg)	92.23 ± 22.69	87.67 ± 13.75	1.11	12.78–3.66	0.27
FBG (mg/dL)	101.65 ± 11.93	97.87 ± 8.97	1.53	8.75–1.19	0.132
PPG(mg/dL)	128.64 ± 11.83	132.64 ± 12.74	−1.25	2.50–10.50	0.21
Hs-CRP (mg/L)	2.95 ± 0.45	4.65 ± 1.84	−4.09	−3.39	0.0006 **

Data presented as mean ± SD, * $p < 0.05$, ** $p < 0.01$ versus control; cTn-I: cardiac troponin-I; TG: triglyceride; TC: total cholesterol; LDL: low density lipoprotein; HDL: high density lipoprotein; VLDL: very low density lipoprotein; AI: atherogenic index; CRR: cardiac risk ratio; SBP: systolic blood pressure; DBP: diastolic blood pressure; FBG: fasting blood glucose; PPG: postprandial glucose; Hs-CRP: highly sensitive CRP.

Regarding the intra-group and inter-group differences, unstable angina versus control showed a significant difference in vaspin and cardiac troponin-I sera levels in addition to the other cardio-metabolic variables, except in cardiac risk ratio (CCR); in the same manner, these differences were found between STEMI and NSTEMI versus control. In the rosuvastatin-treated group, patients with STEMI showed higher vaspin serum levels and lower atherogenic index compared to NSTEMI Table 3.

Table 3. Intra and inter-groups difference in cardio-metabolic variables.

Cardio-Metabolic Variables	STEMI (*n* = 25)	NSTEMI (*n* = 25)	Unstable Angina (*n* = 20)	Control (*n* = 40)
Serum cTn-I (pg/mL)	74.12 ± 12.45	72.65 ± 22.49	54.45 ± 23.53 **	77.64 ± 12.22
Serum vaspin (pg/mL)	611.32 ± 33.64 **	588.67 ± 22.29 **	586.87 ± 33.72 **	542.75 ± 38.95
TC (mg/dL)	197.28 ± 21.77 **	192.99 ± 19.64 **	193.64 ± 20.65 **	266.43 ± 16.59
TG (mg/dL)	160.83 ± 19.42 **	167.54 ± 18.74 **	166.83 ± 16.84 **	254.73 ± 22.82
LDL (mg/dL)	117.78 ± 8.75 **	113.85 ± 9.98 **	112.63 ± 13.83 **	164.71 ± 13.59
HDL (mg/dL)	47.33 ± 9.77	45.63 ± 8.99	45.55 ± 8.33 *	31.76 ± 7.34
VLDL (mg/dL)	32.16 ± 5.88 **	33.50 ± 4.34 **	33.57 ± 4.22 **	50.94 ± 6.42
AI	0.171 ± 0.011 **	0.205 ± 0.014 **	0.204 ± 0.012 **	0.341 ± 0.021
CRR	4.16 ± 1.65	4.22 ± 1.99	4.25 ± 1.86	5.24 ± 1.98
SBP (mmHg)	162.66 ± 20.33	165.62 ± 19.82	164.76 ± 18.93	155.87 ± 19.63
DBP (mmHg)	91.20 ± 20.39	93.53 ± 20.37	93.77 ± 20.53	87.67 ± 13.75
Pulse pressure (mmHg)	71.46 ± 11.76	72.09 ± 10.74	70.99 ± 11.55	68.20 ± 9.33
FBG (mg/dL)	101.44 ± 10.98	100.54 ± 10.71	104.61 ± 9.88	97.87 ± 8.97
PPG (mg/dL)	129.60 ± 10.82	127.55 ± 9.76	126.83 ± 9.38	132.64 ± 12.74
Hs-CRP (mg/L)	2.93 ± 0.44 **	2.89 ± 0.45 **	1.44 ± 0.11 **	4.65 ± 1.84

Data presented as mean ± SD, * $p < 0.05$, ** $p < 0.01$ versus control, $p < 0.01$ (STEMI versus NSTEMI); cTn-I: cardiac troponin-I; TG: triglyceride; TC: total cholesterol; LDL: low density lipoprotein; HDL: high density lipoprotein; VLDL: very low density lipoprotein; AI: atherogenic index; CRR: cardiac risk ratio; SBP: systolic blood pressure; DBP: diastolic blood pressure; FBG: fasting blood glucose; PPG: postprandial glucose; Hs-CRP: highly sensitive CRP.

Therefore, vaspin serum levels were higher in rosuvastatin-treated patients with acute coronary syndrome compared to the patients with acute coronary syndrome not treated with rosuvastatin. Additionally, in the rosuvastatin-treated group, patients with STEMI showed higher vaspin serum levels compared to NSTEMI.

Indeed, there is a non-significant negative correlation between vaspin serum levels and Hs-CRP in ACS in rosuvastatin-treated patients Figure 1.

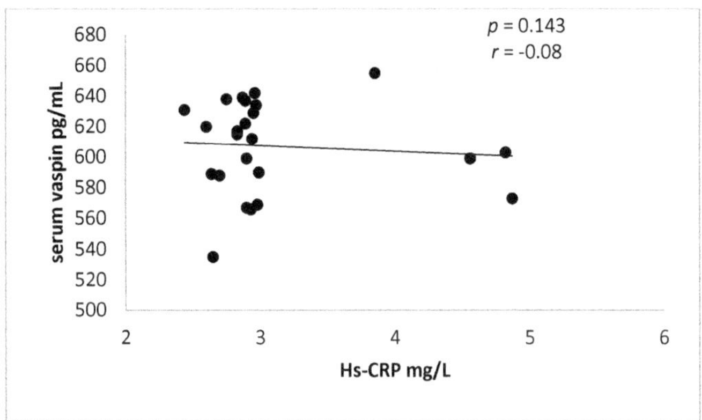

Figure 1. Negative correlation between vaspin serum levels and Hs-CRP in rosuvastatin-treated ACS.

Meanwhile, in control patients (rosuvastatin free), there was a significant negative correlation between vaspin serum levels and Hs-CRP in ACS, see Figure 2.

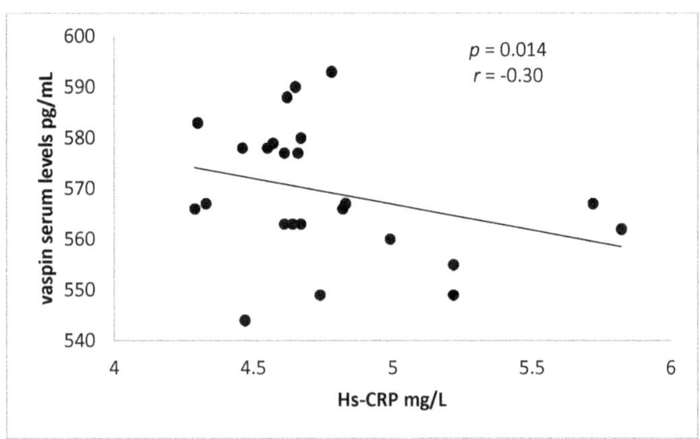

Figure 2. Negative correlation between vaspin serum levels and Hs-CRP in rosuvastatin-free ACS patients.

Vaspin serum levels were positively correlated with cTn-I ($p < 0.001$) and HDL ($p = 0.0003$), but negatively correlated with total cholesterol ($p = 0.007$), triglyceride ($p = 0.04$) and atherogenic index ($p = 0.04$) in patients with STEMI. Meanwhile, in patients with NSTEMI, vaspin serum levels were mainly correlated with cTn-I, total cholesterol and HDL $p < 0.01$; while in patients with unstable angina, vaspin serum levels were only correlated with total cholesterol and HDL $p < 0.01$. In the control patients (rosuvastatin-free patients), vaspin serum levels were correlated with cardio-metabolic risk profile $p < 0.05$, see Table 4.

Regarding the sensitivity and specificity of vaspin serum levels in patients with ACS, high vaspin serum levels(more than 550 pg/mL) were found in 65 patients (treated group) compared to 6 patients (control group) $p < 0.05$ (positive predictive value 0.915 with 95% CI 0.818−0.965), whereas low vaspin serum levels (less than 550 pg/mL) were found in 5 patients (treated group) compared to 34 patients (control group) $p < 0.05$ (negative predictive value 0.871 with 95% CI 0.717−0.951), see Figure 3.

Table 4. Pearson correlation for vaspin serum levels with the cardio-metabolic risk factors in patients with acute coronary syndrome.

Variables	STEMI ($n = 25$)		NSTEMI ($n = 25$)		UA ($n = 20$)		Control ($n = 40$)	
	r	*p*	*r*	*p*	*r*	*p*	*r*	*p*
Serum cTn-I (pg/mL)	0.87	0.001 *	0.77	0.0001 *	0.44	NS	0.94	0.0001 *
TC (mg/dL)	−0.52	0.007 *	−0.67	0.0002 *	−0.67	0.001 *	−0.56	0.007 *
TG (mg/dL)	−0.41	0.04 !	−0.32	NS	−0.36	NS	−0.75	0.001 *
LDL (mg/dL)	−0.33	NS	−0.32	NS	0.38	NS	−0.63	0.0002 *
HDL (mg/dL)	0.66	0.0003 *	0.69	0.0002 *	0.71	0.0004 *	0.51	0.008 *
VLDL (mg/dL)	−0.31	NS	−0.34	NS	−0.38	NS	−0.42	0.006 *
AI	−0.41	0.04 !	−0.35	NS	−0.41	NS	−0.51	0.0007 *
CRR	−0.34	NS	−0.31	NS	−0.37	NS	−0.41	0.008 *
SBP (mmHg)	−0.33	NS	−0.36	NS	−0.32	NS	−0.42	0.006 *
DBP (mmHg)	−0.37	NS	−0.30	NS	−0.22	NS	−0.41	0.008 *
FBG (mg/dL)	−0.28	NS	−0.32	NS	−0.30	NS	−0.33	0.03 !

Pearson correlation (r); * $p < 0.01$; ! $p < 0.05$; NS: non-significant; UA: unstable angina; cTn-I: cardiac troponin-I; TG: triglyceride; TC: total cholesterol; LDL: low density lipoprotein; HDL: high density lipoprotein; VLDL: very low density lipoprotein; AI: atherogenic index; CRR: cardiac risk ratio; SBP: systolic blood pressure; DBP: diastolic blood pressure; FBG: fasting blood glucose.

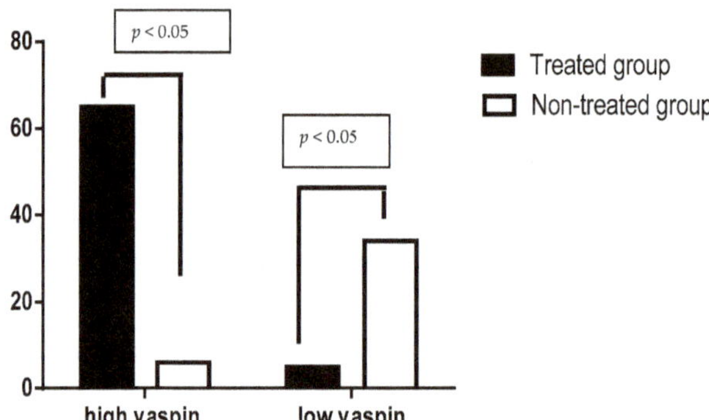

Figure 3. Sensitivity and specificity of vaspin serum levels in patients with acute coronary syndrome treated with or without rosuvastatin.

Thus, vaspin serum levels were high in patients with ACS on rosuvastatin therapy (595.62 ± 23.65) compared to patients with ACS not on rosuvastatin therapy (542.75 ± 38.95), see Figure 4.

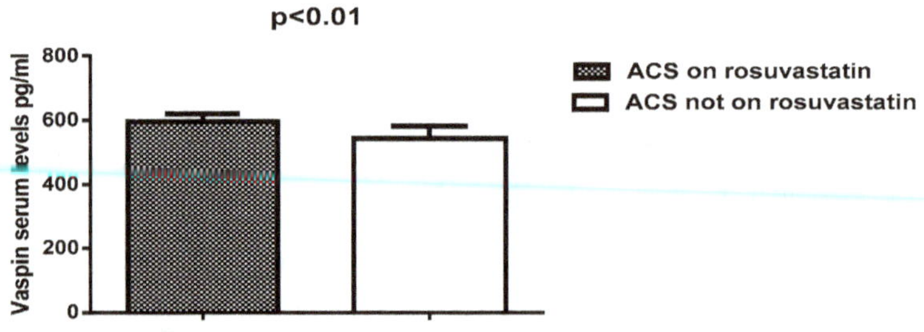

Figure 4. Vaspin serum levels in patients with ACS regarding rosuvastatin therapy.

6. Discussion

Vaspin was initially recognized as an adipokine, mainly secreted from visceral adipose tissue in Otsuka Long-Evans Tokushima Fatty (OLETF), which is an animal model of obesity and diabetes mellitus [20]. Higher vaspin mRNA expression is correlated with type 2 diabetes mellitus, obesity and insulin resistance; this increment in vaspin serum levels leads to improvements in glucose metabolism and insulin sensitivity in visceral adipose tissue, but the main molecular mechanism of vaspin is unidentified [21]. Moreover, vaspin plays an important role in the modulation of coronary vessel homeostasis, since vaspin has a local effect on coronary vascular endothelium [22].

The present study demonstrated higher vaspin serum levels in both the study and control groups; given that the patients in both groups were obese class I type (BMI > 30 kg/m^2), this corresponds to the study by Cho et al., which revealed a significant correlation between vaspin levels and BMI, since vaspin concentrations were increased and correlated with elevated total body fat percentage. For this reason, vaspin concentrations can be regarded as an indicator of obesity [23]. The waist-hip ratio of enrolled patients in the present study was high in both enrolled men and women. This ratio was associated with a high vaspin concentration; this finding correlates with that of the Amouzad et al., 2014 study, which revealed a positive link between elevated waist-hip ratio and serum vaspin concentrations in obesity and metabolic syndromes [24]. Contradictory studies have reported low vaspin serum concentration in obesity compared to normal healthy volunteers [25], which does not correspond with the findings of the present study.

Serum vaspin concentrations of our work were relatively low in patients with ACS, which is consistent with the study by Kobat et al., which demonstrated low vaspin serum levels in patients with ACS and coronary atherosclerosis [26].

The current study also revealed a higher vaspin serum concentration in patients with ACS previously and currently treated with rosuvastatin compared with patients with ACS not previously and currently treated by statins; this result is supported by the 2011 study by Kadoglou et al., which revealed that the pleiotropic effects of statin may increase vaspin serum levels in patients with ischemic heart disease [27].

Moreover, serum vaspin concentrations may be equal in unstable angina and NSTEMI, but there was a significant difference in serum vaspin concentrations between NSTEMI and STEMI. When it is higher in STEMI, this may be due to rosuvastatin effects, and all of the enrolled patients were obese since vaspin serum levels are higher in obesity and are provoked by statins therapy, because statins produced more significant anti-inflammatory effects in STEMI than NSTEMI [28].

Rosuvastatin has potent immuno-modulatory and anti-inflammatory effects due to its inhibition of pro-inflammatory mediators, including tumor necrosis factor (TNF-α), interleukin-6 and C-reactive protein [29], which corresponds with our findings, which revealed a decline in highly sensitive C-reactive protein (hs-CRP) in rosuvastatin-treated patients compared to non-treated patients with ACS. This may explain the decremental effects of rosuvastatin on morbidity of cardio-metabolic risk profile independent of/from lipid-lowering property, which is called the pleiotropic effect [30].

Consequently, the 2011 study of Kodaglou et al. pointed out that inflammatory pathways are the link between obesity and coronary heart diseases, vaspin serum levels showed an inverse association with acute cardiac ischemic events, so low vaspin concentrations were correlated with ACS severity, which is suggestive of the cardio-protective effects of vaspin [31].

Vaspin plays an important role in the prevention of vascular and coronary endothelial injuries and inflammation through down-regulation of intracellular adhesion molecules induced by TNF-α, inhibition of reactive oxygen species, inhibition of platelet-derived growth factor, and suppression of free radical-dependent p38/HSP27 activation [32,33]. Thus, rosuvastatin, like other statins, leads to direct or indirect anti-inflammatory effect via increment in vaspin serum levels that ameliorates acute inflammatory changes in ACS [34], as presented in the present study through elevation of vaspin serum levels and decrement in Hs-CRP.

Additionally, vaspin shows a diurnal variation, high at morning and fasting, and low at the postprandial period; additionally, high vaspin serum levels were reported to be higher in the Asian population [35], these variations may explain high vaspin serum levels in the current study, because all of the selected patients were Asians, and blood sampling was done in the morning to exclude these variations.

Indeed, hs-CRP serum level is an inflammatory marker that increased within two days following ACS, but not in unstable angina; this increment continued for three months subsequent to myocardial infarction [36]. Rosuvastatin significantly reduces Hs-CRP serum levels through its anti-inflammatory properties, as shown in the present study.

Finally, the present study illustrated the inverse correlation of vaspin serum levels with most cardio-metabolic risk profiles in patients with acute coronary syndromes not treated with rosuvastatin, compared to rosuvastatin-treated patients. These findings are in agreement with many studies showing that vaspin levels were found to have an inverse link with the cardio-metabolic events, signifying the protective effect of vaspin in the prevention of coronary atherosclerosis and amelioration of cardiac risk factors [37,38]. Unfortunately, this study did not measure insulin levels; additionally, gender and race differences were not evaluated. These limitations may be a project for future research.

7. Conclusions

Rosuvastatin significantly increases vaspin serum levels in acute coronary syndrome.

Author Contributions: Hayder M. Al-Kuraishy conceived and designed the study experiment. Ali I. Al-Gareeb performed the experiments and analyzed the data. Ali K.Al-Buhadilly contributed reagents/ materials/analysis. All authors contributed in wrote of the paper.

Conflicts of Interest: The authors declare no conflict of interest.

References

1. Tsaousi, G.; Pitsis, A.A.; Ioannidis, G.D.; Vasilakos, D.G. A multidisciplinary approach to unplanned conversion from off-pump to on-pump beating heart coronary artery revascularization in patients with compromised left ventricular function. *Crit. Care Res. Pract.* **2014**, *2014*, 348021. [CrossRef] [PubMed]
2. Erne, P.; Radovanovic, D.; Seifert, B.; Bertel, O.; Urban, P.; AMIS Plus Investigators. The outcome of patients admitted with acute coronary syndrome on palliative treatment: Insights from the nationwide AMIS Plus Registry 1997–2014. *BMJ Open* **2015**, *5*, e006218. [CrossRef] [PubMed]
3. Hicks, K.; Cocks, K.; Corbacho Martin, B.; Elton, P.; MacNab, A.; Colecliffe, W.; Furze, G. An intervention to reassure patients about test results in rapid access chest pain clinic: A pilot randomised controlled trial. *BMC Cardiovasc. Disord.* **2014**, *14*, 138. [CrossRef] [PubMed]
4. Al-Kuraishy, H.M.; Al-Gareeb, A.I.; Awad, M.S.; Alrifai, S.B. Assessment of serum prolactin levels in acute myocardial infarction: The role of pharmacotherapy. *Indian J. Endocrinol. Metab.* **2016**, *20*, 72–79. [CrossRef] [PubMed]
5. Al-Kuraishy, H.M.; Al-Gareeb, A.I. New Insights into the Role of Metformin Effects on Serum Omentin1 Levels in Acute MyocardialInfarction: Cross-Sectional Study. *Emerg. Med. Int.* **2015**, *2015*, 283021.
6. Meyer-Saraei, R.; de Waha, S.; Eitel, I.; Desch, S.; Scheller, B.; Böhm, M.; Lauer, B.; Gawaz, M.; Geisler, T.; Gunkel, O.; et al. Thrombus aspiration in non-ST-elevation myocardial infarction—The 12-month clinical outcome of the randomised TATORT-NSTEMI trial. *Eur. Heart J. Acute Cardiovasc. Care* **2015**, *18*, 10–17. [CrossRef] [PubMed]
7. Ndrepepa, G.; Braun, S.; Tada, T.; Guerra, E.; Schunkert, H.; Laugwitz, K.L.; Kastrati, A. Comparative prognostic value of low-density lipoprotein cholesterol and C-reactive protein in patients with stable coronary artery disease treated with percutaneous coronary intervention and chronic statin therapy. *Cardiovasc. Revasc. Med.* **2014**, *15*, 131–136. [CrossRef] [PubMed]
8. Ndrepepa, G.; Braun, S.; Tada, T.; King, L.; Cassese, S.; Fusaro, M.; Keta, D.; Kastrati, A.; Schmidt, R. Comparative prognostic value of C-reactive protein & fibrinogen in patients with coronary artery disease. *Indian J. Med. Res.* **2014**, *140*, 392–400. [PubMed]

9. Xia, J.; Qu, Y.; Yin, C.; Xu, D. Preoperative rosuvastatin protects patients with coronary artery disease undergoing noncardiac surgery. *Cardiology* **2015**, *131*, 30–37. [CrossRef] [PubMed]

10. Esaki, E.; Adachi, H.; Hirai, Y.; Yamagishi, S.; Kakuma, T.; Enomoto, M.; Fukami, A.; Kumagai, E.; Ohbu, K.; Obuchi, A.; et al. Serum vaspin levels are positively associated with carotid atherosclerosis in a general population. *Atherosclerosis* **2014**, *233*, 248–252. [CrossRef] [PubMed]

11. Shaker, O.G.; Sadik, N.A. Vaspin gene in rat adipose tissue: Relation to obesity-induced insulin resistance. *Mol. Cell. Biochem.* **2013**, *373*, 229–239. [CrossRef] [PubMed]

12. DeClercq, V.; Enns, J.E.; Yeganeh, A.; Taylor, C.G.; Zahradka, P. Modulation of cardiovascular function by adipokines. *Cardiovasc. Hematol. Disord. Drug Targets* **2013**, *13*, 59–72. [CrossRef] [PubMed]

13. Choi, B.J.; Matsuo, Y.; Aoki, T.; Kwon, T.G.; Prasad, A.; Gulati, R.; Lennon, R.J.; Lerman, L.O.; Lerman, A. Coronary endothelial dysfunction is associated with inflammation and vasa vasorum proliferation in patients with early atherosclerosis. *Arterioscler. Thromb. Vasc. Biol.* **2014**, *34*, 2473–2477. [CrossRef] [PubMed]

14. Moradi, S.; Mirzaei, K.; Abdurahman, A.A.; Keshavarz, S.A.; Hossein-Nezhad, A. Mediatory effect of circulating vaspin on resting metabolic rate in obese individuals. *Eur. J. Nutr.* **2015**, *10*, 1–9. [CrossRef] [PubMed]

15. Kêkê, L.M.; Samouda, H.; Jacobs, J.; Di Pompeo, C.; Lemdani, M.; Hubert, H.; Zitouni, D.; Guinhouya, B.C. Body mass index and childhood obesity classification systems: A comparison of the French, International Obesity Task Force (IOTF) and World Health Organization (WHO) references. *Rev. Epidemiol. Sante Publique* **2015**, *63*, 173–182. [CrossRef] [PubMed]

16. Abeles, A.M.; Pillinger, M.H. Statins as antiinflammatory and immunomodulatory agents: A future in rheumatologic therapy? *Arthritis Rheumatol.* **2006**, *54*, 393–407. [CrossRef] [PubMed]

17. Al-Kuraishy, H.M.; Al-Gareeb, A.I. Effects of Rosuvastatin Alone or in Combination withOmega3 FattyAcid on Adiponectin Levelsand Cardiometabolic Profile. *J. Basic Clin. Pharm.* **2016**, *8*, 8–14. [CrossRef] [PubMed]

18. Abudukeremu, N.; Pan, S.; Ma, Y.; Yang, Y.; Ma, X.; Li, X.; Fu, Z.; Huang, Y.; Xie, X.; Liu, F.; et al. The optimal cutoff point of the waist-to-hip ratio for screening Uyghur population aged 35 years and over at high-risk of cardiovascular diseases in Xinjiang. *Zhonghua Xin Xue Guan Bing Za Zhi* **2015**, *43*, 173–178. [PubMed]

19. Al-Kuraishy, H.M.; Al-Gareeb, A.I. Acylation-stimulating protein is a surrogate biomarker for acute myocardial infarction: Role of statins. *J. Lab. Physicians* **2017**, *9*, 163–169. [CrossRef] [PubMed]

20. Blüher, M. Vaspin in obesity and diabetes: Pathophysiological and clinical significance. *Endocrine* **2012**, *41*, 176–182. [CrossRef] [PubMed]

21. Heiker, J.T.; Klöting, N.; Kovacs, P.; Kuettner, E.B.; Sträter, N.; Schultz, S.; Kern, M.; Stumvoll, M.; Blüher, M.; Beck-Sickinger, A.G.; et al. Vaspin inhibits kallikrein 7 by serpin mechanism. *Cell. Mol. Life Sci.* **2013**, *70*, 2569–2583. [CrossRef] [PubMed]

22. Wozniak, S.E.; Gee, L.L.; Wachtel, M.S.; Frezza, E.E. Adipose tissue: The new endocrine organ? A review article. *Dig. Dis. Sci.* **2009**, *54*, 1847–1856. [CrossRef] [PubMed]

23. Cho, J.K.; Han, T.K.; Kang, H.S. Combined effects of body mass index and cardio/respiratory fitness on serum vaspin concentrations in Korean young men. *Eur. J. Appl. Physiol.* **2010**, *108*, 347–353. [CrossRef] [PubMed]

24. Mahdirejei, H.A.; Abadei, S.F.R.; Seidi, A.A.; Gorji, N.E.; Kafshgari, H.R.; Pour, M.E.; Khalili, H.B.; Hajeizad, F.; Khayeri, M. Effects of an eight-week resistance training on plasma vaspin concentrations, metabolic parameters levels and physical fitness in patients with type 2 diabetes. *Cell J.* **2014**, *16*, 367–374.

25. Kim, J.M.; Kim, T.N.; Won, J.C. Association between serum vaspin level and metabolic syndrome in healthy Korean subjects. *Metab. Syndr. Relat. Disord.* **2013**, *11*, 385–391. [CrossRef] [PubMed]

26. Kobat, M.A.; Celik, A.; Balin, M.; Altas, Y.; Baydas, A.; Bulut, M.; Aydin, S.; Dagli, N.; Yavuzkir, M.F.; Ilhan, S. The investigation of serum vaspin level in atherosclerotic coronary artery disease. *J. Clin. Med. Res.* **2012**, *4*, 110–113. [CrossRef] [PubMed]

27. Kadoglou, N.P.; Vrabas, I.S.; Kapelouzou, A.; Lampropoulos, S.; Sailer, N.; Kostakis, A.; Liapis, C.D. Impact of atorvastatin on serum vaspin levels in hypercholesterolemic patients with moderate cardiovascular risk. *Regul. Pept.* **2011**, *170*, 57–61. [CrossRef] [PubMed]

28. Satoh, M.; Takahashi, Y.; Tabuchi, T.; Minami, Y.; Tamada, M.; Takahashi, K.; Itoh, T.; Morino, Y.; Nakamura, M. Cellular and molecular mechanisms of statins: An update on pleiotropic effects. *Clin. Sci.* **2015**, *129*, 93–105. [CrossRef] [PubMed]

29. Vidyarthi, M.; Jacob, P.; Chowdhury, T.A. Oral use of "Low and Slow" Rosuvastatin with Co-Enzyme Q10 in patients with Statin-Induced Myalgia: Retrospective case review. *Indian J. Endocrinol. Metab.* **2012**, *16* (Suppl. S2), S498–S500. [PubMed]

30. Kavalipati, N.; Shah, J.; Ramakrishan, A.; Vasnawala, H. Pleiotropic effects of statins. *Indian J. Endocrinol. Metab.* **2015**, *19*, 554–562. [PubMed]

31. Kadoglou, N.P.; Gkontopoulos, A.; Kapelouzou, A.; Fotiadis, G.; Theofilogiannakos, E.K.; Kottas, G.; Lampropoulos, S. Serum levels of vaspin and visfatin in patients with the coronary artery disease-Kozani study. *Clin. Chim. Acta* **2011**, *412*, 48–52. [CrossRef] [PubMed]

32. Phalitakul, S.; Okada, M.; Hara, Y.; Yamawaki, H. Vaspin prevents TNF-α-induced intracellular adhesion molecule-1 via inhibiting reactive oxygen species-dependent NF-κB and PKC activation in cultured rat vascular smooth muscle cells. *Pharmacol. Res.* **2011**, *64*, 493–500. [CrossRef] [PubMed]

33. Phalitakul, S.; Okada, M.; Hara, Y.; Yamawaki, H. A novel adipocytokine, vaspin inhibits platelet-derived growth factor-BB-induced migration of vascular smooth muscle cells. *Biochem. Biophys. Res. Commun.* **2012**, *423*, 844–849. [CrossRef] [PubMed]

34. Al-Azzam, S.I.; Alzoubi, K.H.; Abeeleh, J.A.; Mhaidat, N.M.; Abu-Abeeleh, M. Effect of statin therapy on vaspin levels in type 2 diabetic patients. *Clin. Pharmacol.* **2013**, *5*, 33–38. [CrossRef] [PubMed]

35. Teshigawara, S.; Wada, J.; Hida, K.; Nakatsuka, A.; Eguchi, J.; Murakami, K.; Kanzaki, M.; Inoue, K.; Terami, T.; Katayama, A.; et al. Serum vaspin concentrations are closely related to insulin resistance, with higher serum levels in Japanese population. *J. Clin. Endocrinol. Metab.* **2012**, *97*, E1202–E1207. [CrossRef] [PubMed]

36. Miyashita, H.; Honda, T.; Maekawa, T.; Takahashi, N.; Aoki, Y.; Nakajima, T.; Tabeta, K.; Yamazaki, K. Relationship between serum antibody titers to Porphyromonas gingivalis and hs-CRP levels as inflammatory markers of periodontitis. *Arch. Oral Biol.* **2012**, *57*, 820–829. [CrossRef] [PubMed]

37. Motawi, T.M.K.; Mahdy, S.G.; El-Sawalhi, M.M.; Ali, E.N.; El-Telbany, R.F.A. Serum levels of chemerin, apelin, vaspin, and omentin-1 in obese type 2 diabetic Egyptian patients with coronary artery stenosis. *Can. J. Physiol. Pharmacol.* **2017**, *28*, 1–7. [CrossRef] [PubMed]

38. Sathyaseelan, A.J.; Adole, P.S.; Wyawahare, M.; Saya, R.P. Assessment of Serum VASPIN Levels among Type 2 Diabetes Mellitus Patients with or without Acute Coronary Syndrome. *J. Clin. Diagn. Res.* **2016**, *10*, BC07–BC10. [CrossRef] [PubMed]

Review

The Role of Statins in the Management of Patients Undergoing Coronary Artery Bypass Grafting

Dimitrios Siskos [1] and Konstantinos Tziomalos [2,*]

[1] Department of Cardiothoracic Surgery, Heart Center, University Hospital Cologne, 50937 Cologne, Germany; dimitrios.siskos@uk-koeln.de

[2] First Propedeutic Department of Internal Medicine, Medical School, Aristotle University of Thessaloniki, AHEPA Hospital, 54636 Thessaloniki, Greece

* Correspondence: ktziomalos@yahoo.com; Tel.: +30-2310-994621; Fax: +30-2310-994773

Received: 13 October 2018; Accepted: 8 November 2018; Published: 11 November 2018

Abstract: Each year, a large number of patients undergo coronary artery bypass grafting surgery (CABG) worldwide. Accumulating evidence suggests that the preoperative administration of statins might be useful in preventing adverse events after CABG. In the present review, we discuss the role of statins in the perioperative management of patients undergoing CABG. Preoperative administration of statins in these patients substantially reduces the risk of postoperative atrial fibrillation and shortens hospital and intensive care unit (ICU) stay. Atorvastatin appears to be more effective, particularly when administered at high doses. Given these benefits and the safety of statins, their administration should be considered in patients undergoing CABG, even though the statins do not appear to affect the incidence of cardiovascular events and overall mortality perioperatively.

Keywords: statins; coronary artery bypass grafting; pleiotropic actions; atrial fibrillation; stroke; acute kidney injury

---

1. Introduction

Each year, a large number of patients undergo coronary artery bypass grafting surgery (CABG) worldwide [1]. Despite improved surgical techniques, the advances in technology used in the operating room (new cardiopulmonary bypass machines and cannulas, more precise sensors and monitoring), improved cardiac anesthesiologic management, and the more experience gathered by surgeons over the years, CABG is still associated with a considerable risk for adverse events, including death [1]. As a result, lowering postoperative morbidity is not only an intraoperative challenge, but also depends on preoperative management. In this context, it is worthy to look back over the publications from years 2000–2007, accumulating evidence that suggests that the preoperative administration of statins might be useful in preventing adverse events after CABG [2–4].

In the present review, we discuss the role of statins in the perioperative management of patients undergoing CABG.

2. Effects of Statins on Atrial Fibrillation after CABG

In the Atorvastatin for reduction of myocardial Dysrhythmia after cardiac surgery (ARMYDA-3 study), a randomized, placebo-controlled study ($n = 200$), treatment with atorvastatin (40 mg/day for seven days) before cardiac surgery (not only isolated CABG) reduced the incidence of postoperative atrial fibrillation by 61% [5]. In a meta-analysis of 3 randomized controlled trials (RCTs) and 10 observational studies ($n = 17,643$), statin treatment before CABG reduced the risk for any postoperative atrial fibrillation by 36% and for new-onset atrial fibrillation by 34% [6]. In a more recent meta-analysis of 12 RCTs in 2980 patients, statin treatment before CABG reduced the incidence of postoperative atrial fibrillation by 58% [7]. Notably, atorvastatin reduced the risk by 65%, whereas rosuvastatin was not protective [7]. Rosuvastatin also did not reduce the incidence of post-CABG atrial fibrillation in the Statin Therapy In Cardiac Surgery (STICS) trial, a large RCT (n = 1922) [8]. The reason for the different outcome between atorvastatin and rosuvastatin is mainly unknown. In addition, it should be mentioned that only two small studies evaluated the effects of the administration of pravastatin and simvastatin on the risk of atrial fibrillation in patients undergoing CABG (n = 43 and 44, respectively) [7]. Even though these studies showed no effect of these statins on the incidence of atrial fibrillation, more data are clearly needed to clarify whether these statins might also prevent this complication. In addition, there are no studies that evaluated the effects of other statins (fluvastatin, pitavastatin) on the incidence of atrial fibrillation in patients undergoing CABG [7].

Regarding the association between statin dose and the risk of atrial fibrillation, an early retrospective study showed that a high-dose statin treatment reduces the risk for atrial fibrillation after CABG more than an intermediate-dose treatment, whereas a low-dose treatment has no protective effect [9]. In contrast, in a duration- and dose-response meta-analysis of eight RCTs (n = 774), there was no association between statin dose and risk reduction (p = 0.47) [10]. Nevertheless, it should be emphasized that both the lipid-lowering and the pleiotropic effects of statins (e.g., anti-inflammatory, antioxidant, and antithrombotic actions) as well as the reduction in cardiovascular risk with these agents are clearly dose-dependent [11]. Therefore, it is possible that this meta-analysis included a small number of patients and did not have the statistical power to show an association between statin dose and the incidence of atrial fibrillation. On the other hand, in the former meta-analysis, a longer duration of preoperative statin treatment was associated with a greater reduction in the risk of postoperative atrial fibrillation (3% reduction per day of statin treatment, p = 0.008) [10].

The physiological mechanism(s) underpinning the preventive effect of statin against postoperative atrial fibrillation are still unknown. The pleiotropic effects of statins, including anti-inflammatory, antioxidant, and antithrombotic actions as well as inhibition of neurohormonal activation, appear to play a role [11,12].

3. Effects of Statins on Hospital and Intensive Care Unit (ICU) Stay after CABG

Several studies showed that preoperative statin therapy shortens hospital stay by approximately 0.5 days [7,10,13–15]. Preoperative statin therapy also reduces postoperative ICU stay, but this reduction is more modest (by 3–4 h) [10,13]. It is possible that the reduction in the incidence of atrial fibrillation and its associated complications, including embolic stroke and hemodynamic instability, contribute to the shortening of hospitalization in patients who receive statins prior to CABG [7,10,13–15].

4. Effects of Statins on Renal Failure, Myocardial Infarction, Stroke, and Mortality after CABG

Postoperative acute kidney injury is a frequent complication in patients undergoing CABG [12,13,16–18]. Observational studies suggested that pretreatment with statins reduces the risk of postoperative renal failure and the need for hemodialysis in this population [19–21]. In contrast, an RCT showed that atorvastatin does not affect the incidence of acute kidney injury after CABG [22] whereas in the STICS trial, the rate of postoperative acute kidney injury was higher in patients who received rosuvastatin prior to CABG compared with patients who received a placebo [8]. Several systematic reviews and meta-analyses concluded that treatment with statins before CABG does not affect the risk of acute kidney injury [12,13,18].

Preoperative treatment with statins does not appear to reduce the risk for postoperative myocardial infarction (MI) [13,17]. However, a meta-analysis of eight studies (n = 8676) showed that patients who receive a loading dose of statin prior to CABG have a lower risk for non-fatal and fatal MI as well as a lower risk for graft restenosis and repeat CABG than patients who receive a regular dose [23]. In a prospective observational study, patients who received high-intensity statin treatment (i.e., expected to reduce low-density lipoprotein cholesterol (LDL-C) levels by >45%) had a lower risk for cardiovascular events than patients who received moderate-intensity statin treatment (i.e., expected to reduce LDL-C levels by <45%) [24].

The incidence of postoperative ischemic stroke also does not appear to be affected by the administration of statins before CABG [13,16,17]. However, a single-center study suggested that the combination of statins with beta blockers reduces the risk for ischemic stroke after CABG [25]. Clearly, more studies are needed to confirm this promising finding.

An early retrospective cohort study suggested that pretreatment with statins is associated with reduced postoperative 30-day all-cause mortality after CABG [17]. However, several more recent RCTs and meta-analyses did not identify any survival benefit with statin treatment in these patients [13,15,16].

5. Conclusions

The preoperative administration of statins in patients undergoing CABG substantially reduces the risk of postoperative atrial fibrillation and shortens hospital and ICU stay. Atorvastatin appears to be more effective, particularly when administered at high doses. Given these benefits and the safety of statins, their administration should be considered in patients undergoing CABG, even though the statins do not appear to affect the incidence of cardiovascular events and overall mortality perioperatively. More studies are needed to clarify the mechanisms underpinning these beneficial effects of statins and to define the optimal compound, dose, and duration of treatment.

Funding: This research received no external funding.

Conflicts of Interest: The authors declare no conflict of interest.

References

1. Alexander, J.H.; Smith, P.K. Coronary-Artery Bypass Grafting. *N. Engl. J. Med.* **2016**, *374*, 1954–1964. [CrossRef] [PubMed]
2. Knatterud, G.L.; Rosenberg, Y.; Campeau, L.; Geller, N.L.; Hunninghake, D.B.; Forman, S.A.; Forrester, J.S.; Gobel, F.L.; Herd, J.A.; Hickey, A.; et al. Long-term effects on clinical outcomes of aggressive lowering of low-density lipoprotein cholesterol levels and low-dose anticoagulation in the post coronary artery bypass graft trial. Post CABG Investigators. *Circulation* **2000**, *102*, 157–165. [CrossRef] [PubMed]
3. White, C.W.; Gobel, F.L.; Campeau, L.; Knatterud, G.L.; Forman, S.A.; Forrester, J.S.; Geller, N.L.; Herd, J.A.; Hickey, A.; Hoogwerf, B.J.; et al. Effect of an aggressive lipid-lowering strategy on progression of atherosclerosis in the left main coronary artery from patients in the post coronary artery bypass graft trial. *Circulation* **2001**, *104*, 2660–2665. [CrossRef] [PubMed]
4. Patti, G.; Pasceri, V.; Colonna, G.; Miglionico, M.; Fischetti, D.; Sardella, G.; Montinaro, A.; Di Sciascio, G. Atorvastatin pretreatment improves outcomes in patients with acute coronary syndromes undergoing early percutaneous coronary intervention: Results of the ARMYDA-ACS randomized trial. *J. Am. Coll. Cardiol.* **2007**, *49*, 1272–1278. [CrossRef] [PubMed]
5. Patti, G.; Chello, M.; Candura, D.; Pasceri, V.; D'Ambrosio, A.; Covino, E.; Di Sciascio, G. Randomized trial of atorvastatin for reduction of postoperative atrial fibrillation in patients undergoing cardiac surgery: Results of the ARMYDA-3 (Atorvastatin for reduction of myocardial Dysrhythmia after cardiac surgery) study. *Circulation* **2006**, *114*, 1455–1461. [CrossRef] [PubMed]
6. Liakopoulos, O.J.; Choi, Y.H.; Kuhn, E.W.; Wittwer, T.; Borys, M.; Madershahian, N.; Wassmer, G.; Wahlers, T. Statins for prevention of atrial fibrillation after cardiac surgery: A systematic literature review. *J. Thorac. Cardiovasc. Surg.* **2009**, *138*, 678–686. [CrossRef] [PubMed]
7. Elgendy, I.Y.; Mahmoud, A.; Huo, T.; Beaver, T.M.; Bavry, A.A. Meta-analysis of 12 trials evaluating the effects of statins on decreasing atrial fibrillation after coronary artery bypass grafting. *Am. J. Cardiol.* **2015**, *115*, 1523–1528. [CrossRef] [PubMed]
8. Zheng, Z.; Jayaram, R.; Jiang, L.; Emberson, J.; Zhao, Y.; Li, Q.; Du, J.; Guarguagli, S.; Hill, M.; Chen, Z.; et al. Perioperative Rosuvastatin in cardiac surgery. *N. Engl. J. Med.* **2016**, *374*, 1744–1753. [CrossRef] [PubMed]

9.   Kourliouros, A.; De Souza, A.; Roberts, N.; Marciniak, A.; Tsiouris, A.; Valencia, O.; Camm, J.; Jahangiri, M. Dose-related effect of statins on atrial fibrillation after cardiac surgery. *Ann. Thorac. Surg.* **2008**, *85*, 1515–1520. [CrossRef] [PubMed]

10. Chen, W.T.; Krishnan, G.M.; Sood, N.; Kluger, J.; Coleman, C.I. Effect of statins on atrial fibrillation after cardiac surgery: A duration- and dose-response meta-analysis. *J. Thorac. Cardiovasc Surg.* **2010**, *140*, 364–372. [CrossRef] [PubMed]

11. Athyros, V.G.; Kakafika, A.I.; Tziomalos, K.; Karagiannis, A.; Mikhailidis, D.P. Pleiotropic effects of statins-clinical evidence. *Curr. Pharm. Des.* **2009**, *15*, 479–489. [CrossRef] [PubMed]

12. Lewicki, M.; Ng, I.; Schneider, A.G. HMG CoA reductase inhibitors (statins) for preventing acute kidney injury after surgical procedures requiring cardiac bypass. *Cochrane Database Syst. Rev.* **2015**, *3*. [CrossRef] [PubMed]

13. Kuhn, E.W.; Slottosch, I.; Wahlers, T.; Liakopoulos, O.J. Preoperative statin therapy for patients undergoing cardiac surgery (Review). *Cochrane Database Syst. Rev.* **2015**, *8*. [CrossRef]

14. Yuan, X.; Du, J.; Liu, Q.; Zhang, L. Defining the role of perioperative statin treatment in patients after cardiac surgery: A meta-analysis and systematic review of 20 randomized controlled trials. *Int. J. Cardiol.* **2017**, *228*, 958–966. [CrossRef] [PubMed]

15. Chopra, V.; Wesorick, D.H.; Sussman, J.B.; Greene, T.; Rogers, M.; Froehlich, J.B.; Eagle, K.A.; Saint, S. Effect of perioperative statins on death, myocardial infarction, atrial fibrillation, and length of stay: A systematic review and meta-analysis. *Arch. Surg.* **2012**, *147*, 181–189. [CrossRef] [PubMed]

16. Thielmann, M.; Neuhäuser, M.; Marr, A.; Jaeger, B.R.; Wendt, D.; Schuetze, B.; Kamler, M.; Massoudy, P.; Erbel, R.; Jakob, H. Lipid-lowering effect of preoperative statin therapy on postoperative major adverse cardiac events after coronary artery bypass surgery. *J. Thorac. Cardiovasc. Surg.* **2007**, *134*, 1143–1149. [CrossRef] [PubMed]

17. Pan, W.; Pintar, T.; Anton, J.; Lee, V.V.; Vaughn, W.K.; Collard, C.D. Statins are associated with a reduced incidence of perioperative mortality after coronary artery bypass graft surgery. *Circulation* **2004**, *110*, 45–49. [CrossRef] [PubMed]

18. Zhao, B.C.; Shen, P.; Liu, K.X. Perioperative statins do not prevent acute kidney injury after cardiac surgery: A meta-analysis of randomized controlled trials. *J. Cardiothorac. Vasc. Anesth.* **2017**, *31*, 2086–2092. [CrossRef] [PubMed]

19. Tabata, M.; Khalpey, Z.; Pirundini, P.A.; Byrne, M.L.; Cohn, L.H.; Rawn, J.D. Renoprotective effect of preoperative statins in coronary artery bypass grafting. *Am. J. Cardiol.* **2007**, *100*, 442–444. [CrossRef] [PubMed]

20. Layton, J.B.; Kshirsagar, A.V.; Simpson, R.J., Jr.; Pate, V.; Jonsson Funk, M.; Stürmer, T.; Brookhart, M.A. Effect of statin use on acute kidney injury risk following coronary artery bypass grafting. *Am. J. Cardiol.* **2013**, *111*, 823–828. [CrossRef] [PubMed]

21. Huffmyer, J.L.; Mauermann, W.J.; Thiele, R.H.; Ma, J.Z.; Nemergut, E.C. Preoperative statin administration is associated with lower mortality and decreased need for postoperative hemodialysis in patients undergoing coronary artery bypass graft surgery. *J. Cardiothorac. Vasc. Anesth.* **2009**, *23*, 468–473. [CrossRef] [PubMed]

22. Billings, F.T., 4th; Hendricks, P.A.; Schildcrout, J.S.; Shi, Y.; Petracek, M.R.; Byrne, J.G.; Brown, N.J. High-Dose Perioperative Atorvastatin and Acute Kidney Injury Following Cardiac Surgery: A Randomized Clinical Trial. *JAMA* **2016**, *315*, 877–888. [CrossRef] [PubMed]

23. Bin, C.; Junsheng, M.; Jianqun, Z.; Ping, B. Meta-analysis of medium and long-term efficacy of loading statins after coronary artery bypass grafting. *Ann. Thorac. Surg.* **2016**, *101*, 990–995. [CrossRef] [PubMed]

24. Ouattara, A.; Benhaoua, H.; Le Manach, Y.; Mabrouk-Zerguini, N.; Itani, O.; Osman, A.; Landi, M.; Riou, B.; Coriat, P. Perioperative statin therapy is associated with significant and dose-dependent reduction of adverse cardiovascular outcomes after coronary artery bypass graft surgery. *J. Cardiothorac. Vasc. Anesth.* **2009**, *23*, 633–638. [CrossRef] [PubMed]

25. Bouchard, D.; Carrier, M.; Demers, P.; Cartier, R.; Pellerin, M.; Perrault, L.P.; Lambert, J. Statin in combination with β-blocker therapy reduces postoperative stroke after coronary artery bypass graft. *Ann. Thorac. Surg.* **2011**, *91*, 654–659. [CrossRef] [PubMed]

 *diseases*

Article

Genomic Influence in the Prevention of Cardiovascular Diseases with a Sterol-Based Treatment

Ismael San Mauro Martín [1,*], Javier Andrés Blumenfeld Olivares [2], Eva Pérez Arruche [2], Esperanza Arce Delgado [2], María José Ciudad Cabañas [3], Elena Garicano Vilar [1] and Luis Collado Yurrita [3]

[1] Research Centers in Nutrition and Health, Paseo de la Habana, 28036 Madrid, Spain; elena@grupocinusa.es
[2] Hospital El Escorial, San Lorenzo de El Escorial, 28200 Madrid, Spain; javierandres.blumenfeld@salud.madrid.org (J.A.B.O.); eva.perez@grupocinusa.es (E.P.A.); e.arcedelgado@gmail.com (E.A.D.)
[3] Department of Medicine, Universidad Complutense de Madrid, 28040 Madrid, Spain; mjciudad@ucm.es (M.J.C.C.); lcollado@med.ucm.es (L.C.Y.)
* Correspondence: info@grupocinusa.es; Tel.: +34-914-577-764

Received: 31 January 2018; Accepted: 28 March 2018; Published: 3 April 2018

Abstract: Raised serum cholesterol concentration is a well-established risk factor in cardiovascular disease. In addition, genetic load may have an indirect influence on cardiovascular risk. Plant-based sterol-supplemented foods are recommended to help reduce the serum low-density lipoprotein cholesterol level. The objective was to analyse the influence of different polymorphisms in hypercholesterolemia patients following a dietary treatment with plant sterols. A randomised double-blind cross-over controlled clinical trial was carried out in 45 people (25 women). Commercial milk, containing 2.24 g of sterols, was ingested daily during a 3-week period, and then the same amount of skim milk, without sterols, was consumed daily during the 3-week placebo phase. Both phases were separated by a washout period of 2 weeks. At the beginning and end of each phase, blood draws were performed. Genes *LIPC C-514T* and *APOA5 C56G* are Ser19Trp carriers and greatly benefit from sterol intake in the diet. *LIPC C-514T* TT homozygous carriers had lower low-density lipoprotein cholesterol (LDL-c) levels than CC homozygote and CT heterozygote carriers after the ingestion of plant sterols ($p = 0.001$). These two genes also showed statistically significant changes in total cholesterol levels ($p = 0.025$; $p = 0.005$), and no significant changes in high-density lipoprotein (HDL) cholesterol levels ($p = 0.032$; $p = 0.003$), respectively. No statistically significant differences were observed for other genes. Further studies are needed to establish which genotype combinations would be the most protective against hypercholesterolemia.

Keywords: genetic; nutrigenetics; sterol; cholesterol; low-density lipoprotein cholesterol; cardiovascular disease

---

1. Introduction

Human health is a result of complex interactions between genetic predisposition and the environment in which genes manifest. It has long been recognised that individual differences in genetic variation influence the association between dietary recommendations and health, which is yet to be reflected in the dietary guidelines. Identifying the interplay between genes and dietary patterns holds promise for a new era of personalised medicine, whereby the recommended diet for best health is tailored towards how an individual's metabolism is genetically predisposed to respond to dietary intake [1].

The understanding of genetic approaches in the nutritional sciences is referred to as nutrigenomics. Nutrigenomics explores the interaction between genetic factors and dietary nutrient intake regarding various disease phenotypes and general health [2], with the aim to provide more personalised dietary advice [3,4].

Through the use of genome-wide association studies, genetic variations (single nucleotide polymorphisms) have been identified as genetic factors, making it more likely to determine individual disease predisposition [2].

Naturally occurring variants of the apolipoprotein A5 gene have been associated with increased triglyceride levels, and have been found to confer risk for cardiovascular diseases. Some of the *MTHFR* gene polymorphisms are also associated with an increased risk of congenital heart failure [5] and cardiovascular disease [6], especially the C677T genetic polymorphism [7]. The hepatic lipase gene (LIPC) is responsible for the hydrolysis of triglycerides [8]. Genetic studies and numerous epidemiologic studies have identified Lp(a) as a risk factor for atherosclerotic diseases, such as coronary heart disease and stroke [9], as it is related to low-density lipoprotein cholesterol (LDL-c) [10]. In addition, the genetic load may have an indirect influence on cardiovascular risk. ApoE plays an essential role in the catabolism of lipoproteins [11].

Cholesterol metabolism is a well-defined responder to dietary intake, and a classic biomarker of cardiovascular health. For this reason, circulating cholesterol levels have become essential in shaping the nutritional recommendations of health authorities worldwide for better management of cardiovascular disease, a leading cause of mortality and one of the most costly health problems globally [12].

A number of cholesterol-related gene–diet interactions have been confirmed, some of which may have clinical importance, supporting a deeper insight into the rapidly emerging field of nutrigenetics for meaningful conclusions that may eventually lead to genetically-targeted dietary recommendations, in the era of personalised nutrition [12].

On the other hand, nutrition and dietary treatments are important in managing cardiovascular disease. Plant sterols have been postulated as beneficial regulatory agents for the control of the disease [13–15]. The daily consumption of phytosterol-enriched foods is widely prescribed as a therapeutic option to reduce LDL-c levels in plasma, and therefore reduce the risk of atherosclerotic disease [16].

Experts from the European Food Safety Authority confirmed that blood cholesterol can be reduced by an average of 7–10.5% if a person consumes 1.5–2.4 g of plant sterols a day. This effect is generally observed within the first 2–3 weeks. Mid- and long-term studies, of up to 85 weeks, show that the blood cholesterol reducing effect could be sustained throughout that period [17].

Future studies should be carried out, with the aim of learning about the increasing difference and soundness of dietary treatment data available in concordance with individual genomic profiles.

The aim of this study was to analyse the influence of different polymorphisms in hypercholesterolemia patients following dietary treatment with plant sterols.

2. Materials and Methods

A randomised double-blind cross-over controlled clinical trial was designed. Participants were recruited at the San Carlos Clinical University Hospital in Madrid (SCCUHM) and the El Escorial Hospital in Madrid. Volunteers throughout the hospital were invited to participate. Patients from primary care and endocrinology departments were specifically invited to participate. Prior to the trial, the participants received instructions on the purpose of the study and signed an informed consent form.

The plant sterols were ingested using Naturcol milk (supplied by *Corporación Alimentaria Peñasanta, S.A.*, Asturias, Spain), available on the market throughout the whole study. Treatment and placebo skim milk were identical in sensorial and nutrient composition.

In this crossover study, all subjects participated in both groups, with two phases of three weeks of each treatment, separated by a two-week washout period (Figure 1). A 2-week washout system was implemented to take into account the metabolism and excretion process, and the 2-day functionality of the sterol in the human body [18].

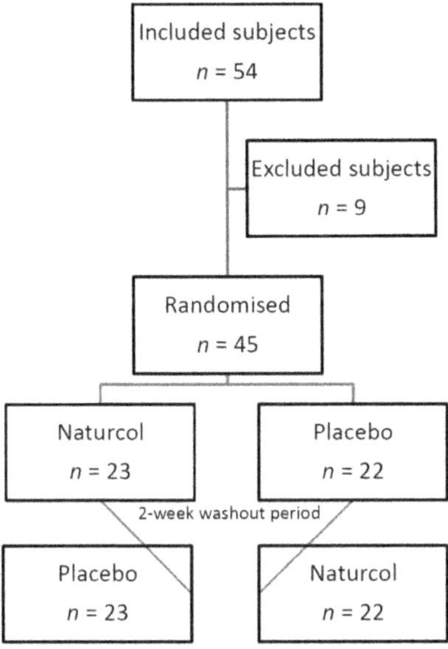

Figure 1. Flowchart of the study participants.

Blood samples were taken at the beginning and at the end of each phase. Over a three-week period, the subjects ingested two glasses of either type of milk daily. Each glass had a standard capacity of 350 mL. Plant sterols were administered daily through the skim milk in quantities of 2.24 g (in plant sterol phase), the same quantity of skim milk without plant sterol was used for the placebo. An 80% target was set as the minimum threshold for consumption. Milk was packaged without labelling to ensure neither the subjects nor the researchers were aware of the type of milk, and were only differentiated by the lid's colour. Groups were randomly assigned, using random number tables.

2.1. Sample Size

Of the 54 initial participants, 45 completed the genetic study: 25 women and 20 men with a mean age of 37.9 $\pm$ 7.5 years. The sample was recruited based on total cholesterol as the primary biomarker, according to the standard formulas for that purpose. The formula was calculated for a difference of $\pm$15% of total cholesterol (200 mg/dL), taking into account a standard deviation of 35 mg/dL for a confidence interval of 95%, a statistical power of 85%, and a loss rate of 10%, using the formula for sample size.

2.2. Inclusion and Exclusion Criteria

Inclusion criteria: men and women, aged 18–50 years, with total cholesterol (TC) > 200 mg/dL.

Exclusion criteria: TC < 200 mg/dL, cardiac pathology (myocardial infarction, angina); lactose intolerance, allergy to cow's milk proteins or plant sterols; obesity (body mass index –BMI– >30) or pharmacological treatment (for cholesterol or fibrate triglycerides, statins, etc.). Subjects that already supplemented with plant sterols. None of the participants presented active thyroid disease, liver disease, alcoholism, or any other such condition that dynamically altered lipid levels and/or diets.

2.3. Clinical Analyses

The samples for the analytical tests were extracted by medical staff after a 12-h fast at the Clinical Analysis Unit in the SCCUHM and El Escorial Hospital, and in the San Carlos Specialty Unit in San Lorenzo de El Escorial.

The blood extraction protocol of the laboratory was as follows: blood extraction was performed with a gel serum tube, using the technique 'Blood collection' with an s-Monovette in aspiration [19–21]. After extraction, the samples were kept at 5 ± 3 °C until their arrival at the laboratory. The centrifugation settings were T 20 ± 5 °C, for 10 minutes, at 1200 g. Stability: 1 week at 5 ± 3 °C. The Ultraviolet–Visible spectrophotometry technique was used.

The laboratory that performed the tests is located in the same city of the study, and has the required legal accreditations, certificates, and standards, ISO 9001:2008, and accreditation 511/LE2114 according to criteria included in the UNE-EN ISO 15189 Standard.

2.4. Genetic

A variation in the effect of dietary treatment with plant sterols in milk on lipoprotein metabolism (reduction of LDL-c) was observed; therefore, a genomic analysis was performed.

With this analysis, we sought to understand the following hypothesis: (1) that genetic polymorphisms could be associated with a higher or lower response effect to the plant sterol treatment for hypercholesterolemia.

Polymorphisms APOA5 C56G Ser19Trp, Prothrombin G20210A, F5 Arg506Gln, MTHFR C677T, LIPC C-514T, LPA I4300M, PPAR_ALPHA L162V, APOA5 1131T>C, APOE APOE2/3/4 and APOE APOE2,3,4 were studied.

Genomic DNA, for genotyping the SNP, was extracted from saliva samples. The analyses were run by Vitagenes Lab (San Francisco, CA, USA). The genotyping was conducted using the Biobank Axiom1 96-Array from Affymetrix. Genotype calling was performed with respect to Affymetrix's best practice guidelines, including analysis with SNPolisher, assuming a quality control rate of >0.97 [22,23].

The extraction and purification of DNA from saliva samples was carried out as follows: each DNA extraction was run on an agarose gel to ensure high quality and high molecular weight. To ensure maximum purity of the extracted DNA (ratio > 1.7), the OD260/280 ratio was analysed, that is, the optical density of the extracted DNA at 260 nm and 280 nm wavelengths. All analyses were performed in duplicate to ensure maximum precision and reliability in their results.

All Vitagenes analytical processes were carried out in laboratories that had CLIA certification, thus attaining the optimal level of technical quality and precision proposed by the FDA in the United States [24].

2.5. Variables and Study Factors

An ad hoc questionnaire was designed for the study. The questionnaire and anthropometric study were performed by a single trained researcher, ensuring the homogeneity and standardisation of the uniformity criteria and methodology to follow. The study variables were established in terms of the proposed objectives: sex, age, clinical and pharmacological history, sleep quality, health habits, use of tobacco and alcohol, intestinal transit, food consumption frequency, and physical activity. In addition, each participant's weight, height, waist perimeter, BMI, fat percentage, subcutaneous fat percentage, and lean body mass (kg) were measured. Weight, BMI and body composition were determined by means of tetrapolar multi-frequency (20 and 100 kHz) electrical bioimpedance, InBody Model 230, following the usual standard protocol and the manufacturer's recommendations [25]. Waist perimeter was measured with a flexible non-elastic metal measuring tape with a range of 0.1 mm–150 cm. Basal metabolic rate was calculated with the following formulas: BMR = 66 + (13.75 × weight in kg) + (5 × height in cm) − (6.8 × age in yrs), for men; BMR = 655 + (9.6 × weight in kg) + (1.8 × height in cm) − (4.7 × age in Yrs), for women. The analytical markers were as follows: lipid profile (total cholesterol, high-density lipoprotein cholesterol (HDL-c), LDL-c, and non-HDL-cholesterol. Confounding factors were also taken into account with an affinity table after ingestion (>95%), monitoring of the non-modification of baseline habits during the trial, and a record of food consumption frequencies; in order to control the ingestion of foods that may influence the metabolism of cholesterol upwards or downwards.

2.6. Statistical Analysis

The data were analysed using the SPSS 21.0 statistical package. A descriptive analysis was first made of the sociodemographic and anthropometric data, the baseline, and the final lipid values under the ingestion of Naturcol and the placebo. The normality of the lipid values was determined using the Shapiro–Wilk test. To analyse the efficacy of the ingestion of Naturcol and the placebo, the difference in lipid values was calculated before and after ingestion, also applying the Student's *T*-test, for related samples, or the Wilcoxon rank sum test according to the compliance with the assumption of normality of the dependent variables. The efficacy of the intervention was verified by comparing the differences (final-baseline) in the ingestion of milk with sterols and the placebo, by applying the Student's *T*-test for related samples, or the Wilcoxon rank sum test depending on the compliance with the assumption of normality of the lipid increase. The effect size and the proportion of the mean differences were calculated with regard to the standard deviation of the baseline, or milk with sterols as the case may be. The level of significance applied was 5%.

The study was approved by the respective bioethics committees of the San Carlos Clinical University Hospital in Madrid (SCCUHM) and the El Escorial Hospital in Madrid.

This study followed the ethical principles enshrined in the Helsinki Declaration, the recommendations for good clinical practice, current Spanish legislation regulating clinical research in humans, and the protection of personal and bioethical data (Royal Decree 561/1993 on clinical trials and 14/2007, 3 July for biomedical research).

3. Results

A total of 45 subjects (25 women and 20 men) completed the trial. Nine subjects failed to complete the trial because of different genotyping, and failure to complete the plant sterols treatment (ingested less than 80% of the consumption protocol). They had an average age of 37.9 ± 7.5 years and weighed 69 kg (BMI 23.7 kg/m^2) (Table 1). Baseline total cholesterol was 236.6 mg/dL, LDL cholesterol was 157.3 mg/dL and HDL cholesterol was 58.2 mg/dL. There were no differences between centres in volunteer demographics. In Table 2, the descriptive statistic of genes and haplotypes can be found.

Table 1. Descriptive statistics of the anthropometric measurements and lipid profile.

	Total ($n = 45$)	Males ($n = 20$)	Females ($n = 25$)
	Mean $\pm$ SD	Mean $\pm$ SD	Mean $\pm$ SD
Age (years)	37.9 ± 7.5	39.7 ± 6.9	36.5 ± 7.8
Weight (kg)	68.6 ± 12.4	77 ± 11.3	61.4 ± 8.03
Height (m)	1.7 ± 0.1	1.76 ± 0.1	1.64 ± 0.1
BMI (Kg/m^2)	23.7 ± 3.1	24.8 ± 3.2	22.8 ± 2.7
Body fat (%)	25.5 ± 7.3	21.2 ± 6.5	29.4 ± 5.7
Visceral fat (kg)	6.8 ± 4.6	8 ± 4.7	5.7 ± 4.3
Muscle (kg)	35.5 ± 13.4	40.4 ± 14.1	31.2 ± 11.4
Basal metabolic rate (kcal)	1484.5 ± 278.6	1705 ± 230.1	1285 ± 125.3
	Decrease % (from baseline to final measures)		
Total cholesterol (mg/dL)	$9.1 + 9.3$	8.2 ± 9.8	9.9 ± 9.3
LDL-cholesterol (mg/dL)	11.8 ± 13.04	9.9 ± 12.9	13.4 ± 13.8
HDL-cholesterol (mg/dL)	1.8 ± 8.1	2.1 ± 6.9	1.5 ± 9.3
Non-HDL-cholesterol (mg/dL)	11.5 ± 10.8	9.6 ± 11.7	13.2 ± 10.3

Table 2. Descriptive statistics of genes and haplotypes.

Gene	Haplotype	Frequency (*n*)	Percentage (%)
APOA5 C56G Ser19Trp (rs3135506)	CG	12	24.5
	GG	37	75.5
Prothrombin G20210A (rs1799963)	GG	49	100
F5 Arg506Gln (rs6025)	GG	49	100
MTHFR C677T (rs1801133)	CC	13	26.5
	CT	31	63.3
	TT	5	10.2
LIPC C-514T (rs1800588)	CC	19	38.8
	CT	24	49
	TT	6	12.2
LPA I4300M (rs3798220)	TT	47	95.9
	TC	2	4.1
PPAR-alpha L162V (rs1800206)	CC	43	87.8
	CG	6	12.2
APOA5 1131T>C (rs662799)	TT	46	93.9
	TC	3	6.1
APOE Haplotipo APOE2/3/4 (rs429358)	TT	42	85.7
	TC	7	14.3
APOE Haplotipo APOE2,3,4 (rs7412)	TC	7	14.3
	CC	42	85.7

No statistically significant differences ($p > 0.05$) were found between the decreased percentage of lipid parameters (total cholesterol, high-density lipoprotein cholesterol (HDL) cholesterol, and Non-HDL cholesterol), and the genotypes observed for the following genes (Figure 2): MTHFR C677T, LPA I4300M, PPAR_ALPHA L162V, APOA5 1131T>C, APOE APOE2/3/4 ($p = 0.416$) and APOE APOE2,3,4.

Only APOA5 C56G Ser19Trp and LIPC C-514T genes incurred statistically significant changes for levels of total cholesterol ($p = 0.025$; $p = 0.005$), but not HDL cholesterol ($p = 0.032$; $p = 0.003$), respectively, after the treatment with sterols (Figure 2).

LIPC C-514T, APOA5 1131T>C and APOE APOE2/3/4 greatly benefit from sterol intake in the diet (Figure 3). In LIPC C-514T carriers, TT homozygous and CT heterozygote carriers lowered their LDL-c more than CC-homozygote carriers, after the ingestion of plant sterols ($p = 0.001$). APOA5 1131T>C and APOE APOE2/3/4 did not show statistically significant differences ($p = 0.682$; $p = 0.416$).

A large difference was also observed in the case of LPA I4300M and PPAR_alpha L162V genes, between the carriers of the homozygous (TT and CC, respectively) and heterozygous (TC and CG, respectively) variants, although the difference is not statistically significant ($p = 0.121$; $p = 0.180$).

Polymorphisms APOA5 C56G Ser19Trp, MTHFR C677T and APOE APOE2,3,4 exert a favourable effect in the treatment, but no statistically significant differences in the reduction of LDL-c can be seen ($p = 0.246$; $p = 0.662$; $p = 0.637$).

The analytical study could not be applied to Prothrombin G20210A and F5 Arg506Gln genes, because each gene had only one haplotype.

After checking whether any genotype was directly related to baseline LDL-c, it was found that only one of the analysed genes (APOA5 C56G Ser19Trp; $p = 0.013$) was associated with different rates of baseline cholesterol, according to their single nucleotide polymorphism (SNP).

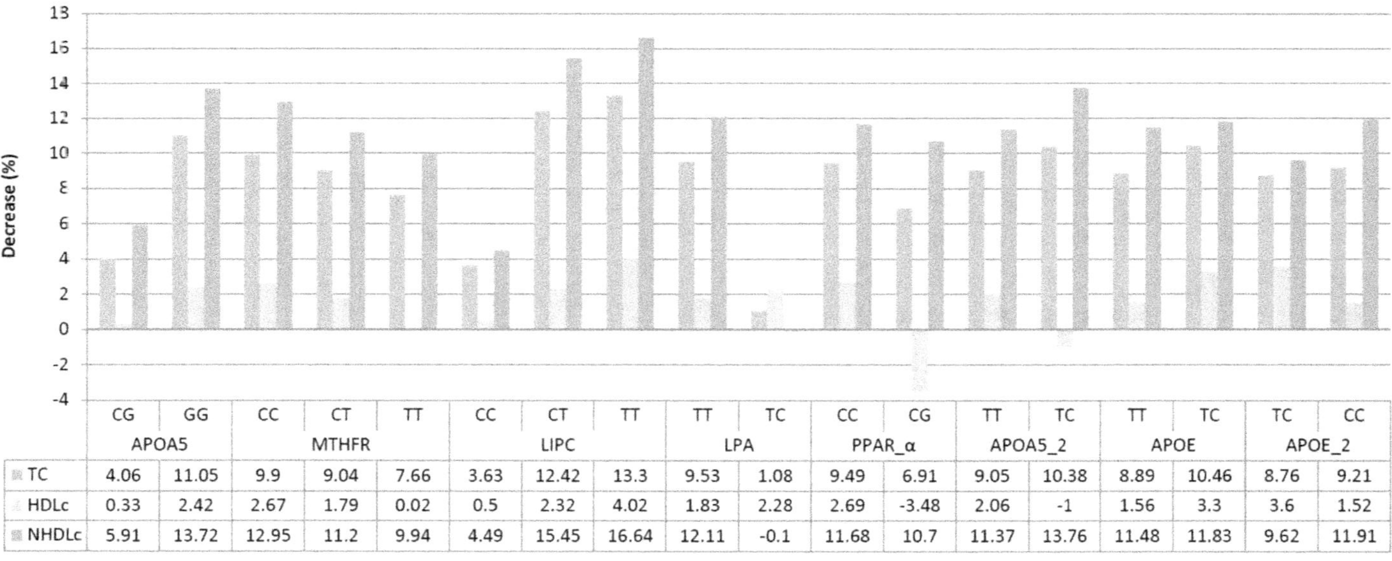

	CG	GG	CC	CT	TT	CC	CT	TT	TT	TC	CC	CG	TT	TC	TT	TC	TC	CC
	APOA5		MTHFR			LIPC			LPA		PPAR_α		APOA5_2		APOE		APOE_2	
TC	4.06	11.05	9.9	9.04	7.66	3.63	12.42	13.3	9.53	1.08	9.49	6.91	9.05	10.38	8.89	10.46	8.76	9.21
HDLc	0.33	2.42	2.67	1.79	0.02	0.5	2.32	4.02	1.83	2.28	2.69	-3.48	2.06	-1	1.56	3.3	3.6	1.52
NHDLc	5.91	13.72	12.95	11.2	9.94	4.49	15.45	16.64	12.11	-0.1	11.68	10.7	11.37	13.76	11.48	11.83	9.62	11.91

Figure 2. Percentage difference in lipid parameters before treatment, based on genes and haplotypes. TC: total cholesterol; HDLc: high density lipoprotein cholesterol; NHDLc: non-high density lipoprotein cholesterol.

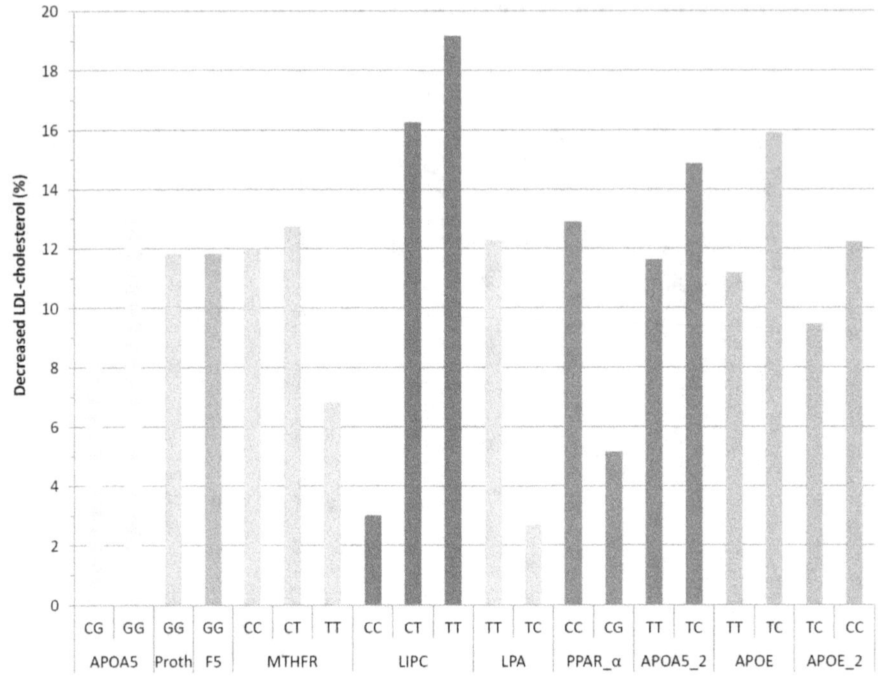

Figure 3. Percentage difference in low-density lipoprotein (LDL) cholesterol values before treatment, according to genes and haplotypes.

Other genes and their haplotypes could not be linked to a higher or lower level of basal cholesterol, as measured by the parametric test, *T*-Student test or ANOVA. This means a possible protection of this risk factor (LDL-c) could not establish by genotype. The obtained results for each gene were as follows: MTHFR C677T ($p = 0.184$), LIPC C-514T ($p = 0.522$), LPA I4300M ($p = 0.158$), PPAR_ALPHA L162V ($p = 0.936$), APOA5 1131T>C ($p = 0.484$), APOE APOE2/3/4 ($p = 0.604$), and APOE APOE2,3,4 ($p = 0.233$).

4. Discussion

Phytosterols are known to reduce serum LDL-c level without changing HDL-c levels. Daily consumption of phytosterol-enriched foods is widely used as a therapeutic option to lower plasma cholesterol and the risk of atherosclerotic disease [26]. Clifton et al. (2004) [27], who analysed the differences of using plant sterols in different matrices, concluded that the matrix that had a better effect in reducing LDL-c was obtained in milk.

When the recent scientific literature is reviewed, important studies [28] evaluating the benefit of plant sterol consumption on cardiovascular risk biomarkers can be found, as a meta-analysis of more than 40 clinical trials concluded [29].

Like other authors [29,30], this study analysed whether the effectiveness of plant sterols is dependent on the subjects' baseline levels; finding a higher decrease in LDL c, the greater baseline, depends on the subjects' innate genes. Lipid metabolism, of great importance regarding cardiovascular risk, is regulated by many genes whose variants may influence this risk [11].

Maasz et al. (2008) [31] suggested that the 56G allele can confer risk exclusively for the development of large-vessel associated stroke. Thereby, the 56G allele differs from the APOA5T-1131C allelic variant, which has been previously identified as a risk factor for all subgroups of the stroke disease, namely the C allele variant, which is a risk factor for heart disease and ischemic stroke [32,33], by increasing levels of triglycerides. This polymorphism has a significant association with coronary heart disease risk [34]. Sanchez-Moreno et al. (2011) [35] found that very low-density lipoprotein cholesterol (VLDL-C) concentrations were higher in carriers of the minor allele than in noncarriers (0.7 ± 0.32 vs. 0.6 ± 0.22 mmol/L) ($p = 0.01$).

Regarding the ApoE, it has been observed that the $\varepsilon4$ allele of the ApoE4 is a major risk factor for coronary heart disease. However, the same results were not observed in the case of ApoE2, whose polymorphisms did not seem significantly related to coronary risk [36].

The relationship between the hepatic lipase gene and the influence on HDL-c has not been clearly established, although it seems to be an inverse relationship [8]. The influence that this gene has on the metabolism of glycerophospholipids [37] may also alter plasma concentrations thereof. The results published by Posadas-Sanchez et al. (2015) [38] suggest that the LIPC C-154T polymorphism is associated with cardiometabolic parameters and cardiovascular risk factors: under the dominant model, the TT genotype was associated with increased levels of the triglycerides/HDL-cholesterol index ($p = 0.046$). On the other hand, the same genotype was associated with the presence of small LDLs ($p = 0.003$). The risk analysis showed that under a dominant model, the LIPCC-514T polymorphism was associated with increased hypertriglyceridemia (OR = 1.36, $p = 0.006$) and coronary artery calcification (OR = 1.44, $p = 0.015$). The T allele carriers had higher levels of low-density lipoprotein cholesterol (LDL-C) in obese boys, and TC and LDL-C in non-obese girls (all $p < 0.05$) [39].

Li et al. (2015) [36] exposed that for the C677T polymorphism, when compared with the wild CC genotype, heterozygosity of CT increased the risk of congenital heart disease (OR = 2.249, 95% CI 1.305–3.877, $p = 0.003$), and the homozygous mutant genotype TT was significantly associated with the risk of congenital heart disease (OR = 3.121, 95% CI 1.612–6.043, $p = 0.001$).

The results suggested that genetic and metabolic biomarkers when used together may predict inter-individual lipid level responsiveness to the plant-sterol-intervention, and thus could be useful in devising individualised cholesterol-lowering strategies. We cannot find an accurate and concise explanation, beyond the observation of these differences in the results, which should undoubtedly be refuted until more research can corroborate our findings, and with samples of the wider population.

However, nutrigenomics has already observed these differences for years, as mutations in some genes are known to hinder gene function, for example, in the classic studies of PUFAS and Framingham [40–42]. It has not supposed a clinical transcendence as suggestive, as revealed by its important results.

5. Conclusions

Having analysed the influence of different polymorphisms in hypercholesterolemia patients following a dietary treatment with plant sterols, we can conclude that gene LIPC C-514T carriers, especially TT homozygous carriers, lowered LDL-c more than CC homozygote and CT heterozygote carriers, after ingestion of plant sterols. LIPC C-514T and APOA5 C56G Ser19Trp carriers greatly improve their lipid profile from sterol intake in the diet. No statistically significant differences were observed for other genes.

Human genome discoveries need to be moved into health practice in a way that maximises health benefits and minimises harm to individuals and populations.

These results provide the basis for further studies to establish which genotype combinations would be the most protective against hypercholesterolemia.

Limitations: The major limitation of this study was the sample size, which limited us from being able to extrapolate the results to the Spanish population. A larger sample size would result in a sophisticated design and a greater budget, that would limit the number of publications aligned in a deeper analysis.

Acknowledgments: We would like to thank *Corporación Alimentaria Peñasanta, S.A.* for their collaboration. We also thank the clinical analysis unit at Hospital Clinico San Carlos and Hospital El Escorial, in Madrid; the Medicine Department of Universidad Complutense de Madrid; Vitagenes and Microcaya S.A.

Author Contributions: I.S.M.M. and J.A.B.O. contributed to the conception and design of the work; E.P.A. and E.A.D. contributed to data collection; E.G.V. contributed to the analysis and interpretation of data, drafting the article; M.J.C.C. and L.C.Y. critically revised the article. All authors have approved the final version for publication.

Conflicts of Interest: The authors declare no conflict of interest. The founding sponsors had no role in the design of the study; in the collection, analyses, or interpretation of data; in the writing of the manuscript, and in the decision to publish the results.

References

1. Frazier-Wood, A.C. Dietary patterns, genes, and health: Challenges and obstacles to be overcome. *Curr. Nutr. Rep.* **2015**, *4*, 82–87. [CrossRef] [PubMed]

2. Doo, M.; Kim, Y. Obesity: Interactions of genome and nutrients intake. *Prev. Nutr. Food Sci.* **2015**, *20*, 1–7. [CrossRef] [PubMed]

3. Corella, D.; Peloso, G.; Arnett, D.K.; Demissie, S.; Cupples, L.A.; Tucker, K.; Lai, C.Q.; Parnell, L.D.; Coltell, O.; Lee, Y.C.; et al. APOA2, dietary fat, and body mass index: Replication of a gene-diet interaction in 3 independent populations. *Arch. Intern. Med.* **2009**, *169*, 1897–1906. [CrossRef] [PubMed]

4. Seip, R.L.; Volek, J.S.; Windemuthl, A.; Kocherla, M.; Fernandez, M.L.; Kraemer, W.J.; Ruaño, G. Physiogenomic comparison of human fat loss in response to diets restrictive of carbohydrate or fat. *Nutr. Metab.* **2008**, *5*, 4. [CrossRef] [PubMed]

5. Chen, K.H.; Chen, L.L.; Li, W.G.; Fang, Y.; Huang, G.Y. Maternal MTHFR C677T polymorphism and congenital heart defect risk in the Chinese Han population: A meta-analysis. *Genet. Mol. Res.* **2013**, *12*, 6212–6219. [CrossRef] [PubMed]

6. Zhang, M.J.; Li, J.C.; Yin, Y.W.; Li, B.H.; Liu, Y.; Liao, S.Q.; Gao, C.Y.; Zhang, L.L. Association of MTHFR C677T polymorphism and risk of cerebrovascular disease in Chinese population: An updated meta-analysis. *J. Neurol.* **2014**, *261*, 925–935. [CrossRef] [PubMed]

7. Kumar, A.; Kumar, P.; Prasad, M.; Sagar, R.; Yadav, A.K.; Pandit, A.K.; Jali, V.P.; Pathak, A. Association of C677T polymorphism in the methylenetetrahydrofolate reductase gene (MTHFR gene) with ischemic stroke: A meta-analysis. *Neurol. Res.* **2015**, *37*, 568–577. [CrossRef] [PubMed]

8. Edmondson, A.C.; Braund, P.S.; Stylianou, I.M.; Khera, A.V.; Nelson, C.P.; Wolfe, M.L.; Derohannessian, S.L.; Keating, B.J.; Qu, L.; He, J.; et al. Dense genotyping of candidate gene loci identifies variants associated with high-density lipoprotein cholesterol. *Circ. Cardiovasc. Genet.* **2011**, *4*, 145–155. [CrossRef] [PubMed]

9. Nordestgaard, B.G.; Chapman, M.J.; Ray, K.; Borén, J.; Andreotti, F.; Watts, G.F.; Ginsberg, H.; Amarenco, P.; Catapano, A.; Descamps, O.S.; et al. Lipoprotein(a) as a cardiovascular risk factor: Current status. *Eur. Heart J.* **2010**, *31*, 2844–2853. [CrossRef] [PubMed]

10. Donnelly, L.A.; Van Zuydam, N.R.; Zhou, K.; Tavendale, R.; Carr, F.; Maitland-Van der Zee, A.H.; Leusink, M.; de Boer, A.; Doevendans, P.A.; Asselbergs, F.W.; et al. Robust association of the LPA locus with low-density lipoprotein cholesterol lowering response to statin treatment in a meta-analysis of 30467 individuals from both randomized control trials and observational studies and association with coronary artery disease outcome during statin treatment. *Pharmacogenet. Genom.* **2013**, *23*, 518–525.

11. San Mauro-Martín, I.; de la Calle-de la Rosa, L.; Sanz-Rojo, S.; Garicano-Vilar, E.; Ciudad-Cabañas, M.J.; Collado-Yurrita, L. Genomic approach to cardiovascular disease. *Nutr. Hosp.* **2016**, *33*, 148–155.

12. Abdullah, M.M.; Jones, P.J.; Eck, P.K. Nutrigenetics of cholesterol metabolism: Observational and dietary intervention studies in the postgenomic era. *Nutr. Rev.* **2015**, *73*, 523–543. [CrossRef] [PubMed]

13. San Mauro Martin, I.; Collado Yurrita, L.; Ciudad Cabañas, M.J.; Cuadrado Cenzual, M.A.; Hernández Cabria, M.; Calle Purón, M.E. Manejo del riesgo de enfermedad cardiovascular con leche enriquecida en esteroles en población joven adulta, ensayo clínico controlado aleatorizado y cruzado. *Nutr. Hosp.* **2014**, *30*, 945–951. [PubMed]

14. National Institutes of Health. Third Report of the National Cholesterol Educcaction Program (NCEP) Expert Panel on Detection, Evaluation, and Treatment of High Blood Cholesterol in Adults (Adult Treatment Panel III). 2002. Available online: http://www.nhlbi.nih.gov/files/docs/resources/heart/atp3full.pdf (accessed on 26 April 2016).

15. Alonso Karlezi, R.A.; Mata Pariente, N.; Mata López, P. Control de las hiperlipemias en la práctica clínica. *Rev. Esp. Cardiol. Supl.* **2006**, *6*, 24–35. [CrossRef]

16. Matthan, N.R.; Zhu, L.; Pencina, M.; D'Agostino, R.B.; Schaefer, E.J.; Lichtenstein, A.H. Sex-specific differences in the predictive value of cholesterol homeostasis markers and 10-year cardiovascular disease event rate in Framingham offspring study participants. *J. Am. Heart Assoc.* **2013**, *2*, e005066. [CrossRef] [PubMed]

17. EFSA Panel on Dietetic Products, Nutrition and Allergies (NDA). Scientific Opinion on the substantiation of health claims related to plant sterols and plant stanols and maintenance of normal blood cholesterol concentrations (ID 549, 550, 567, 713, 1234, 1235, 1466, 1634, 1984, 2909, 3140), and maintenance of normal prostate size and normal urination (ID 714, 1467, 1635) pursuant to Article 13 of Regulation (EC) No 1924/2006. *EFSA J.* **2010**, *8*, 1813.

18. Ling, W.H.; Jones, P.J. Dietary phytosterols: A review of metabolism, benefits and side effects. *Life Sci.* **1995**, *57*, 195–206. [CrossRef]

19. Lippi, G.; Avanzini, P.; Musa, R.; Sandei, F.; Aloe, R.; Cervellin, G. Evaluation of sample hemolysis in blood collected by S-Monovette® using vacuum or aspiration mode. *Biochem. Med. (Zagreb)* **2013**, *23*, 64–69. [CrossRef] [PubMed]

20. Bowen, R.A.R.; Remaley, A.T. Interferences from blood collection tube components on clinical chemistry assays. *Biochem. Med. (Zagreb)* **2014**, *24*, 31–44. [CrossRef] [PubMed]

21. Sarstedt. Blood Collection with the S-Monovette ®. Available online: https://www.sarstedt.com/fileadmin/user_upload/99_Gebrauchsanweisungen/Englisch_US_Code/644_c_PosterA3_AnleitungVenoeseBE_SafetyKanuele_GB_US_0314.pdf (accessed on 8 March 2018).

22. Weale, M.E. Quality control for genome-wide association studies. *Methods Mol. Biol.* **2010**, *628*, 341–372. [PubMed]

23. Ziegler, A. Genome-wide association studies: Quality control and population-based measures. *Genet. Epidemiol.* **2009**, *33*, S45–S50. [CrossRef] [PubMed]

24. Vitagenes. Calidad Tecnica. Available online: https://www.vitagenes.com/Calidad/CalidadTecnica/ (accessed on 12 March 2018).

25. Portao, J.; Bescós, R.; Irurtia, A.; Cacciatori, E.; Vallejo, L. Valoración de la grasa corporal en jóvenes físicamente activos: Antropometría vs. bioimpedancia. *Nutr. Hosp.* **2009**, *24*, 529–534. [PubMed]

26. Rocha, M.; Banuls, C.; Bellod, L.; Jover, A.; Victor, V.M.; Hernandez-Mijares, A. A review on the role of phytosterols: New insights into cardiovascular risk. *Curr. Pharm. Des.* **2011**, *17*, 4061–4075. [CrossRef] [PubMed]

27. Clifton, P.M.; Noakes, M.; Sullivan, D.; Erichsen, N.; Ross, D.; Annison, G.; Fassoulakis, A.; Cehun, M.; Nestel, P. Cholesterol-lowering effects of plant sterol esters differ in milk, yoghurt, bread and cereal. *Eur. J. Clin. Nutr.* **2004**, *58*, 503–509. [CrossRef] [PubMed]

28. Katan, M.B.; Grundy, S.M.; Jones, P.; Law, M.; Miettinen, T.; Paoletti, R. Stresa Workshop Participants. Efficacy and safety of plant stanols and sterols in the management of blood cholesterol levels. *Mayo Clin. Proc.* **2003**, *78*, 965–978. [CrossRef]

29. Abumweis, S.S.; Barake, R.; Jones, P.J. Plant sterols/stanols as cholesterol lowering agents: A meta-analysis of randomized controlled trials. *Food Nutr. Res.* **2008**, *52*. [CrossRef] [PubMed]

30. Naumann, E.; Plat, J.; Kester, A.D.; Mensink, R.P. The baseline serum lipoprotein profile is related to plant stanol induced changes in serum lipoprotein cholesterol and triacylglycerol concentrations. *J. Am. Coll. Nutr.* **2008**, *27*, 117–126. [CrossRef] [PubMed]

31. Maász, A.; Kisfali, P.; Szolnoki, Z.; Hadarits, F.; Melegh, B. Apolipoprotein A5 gene C56G variant confers risk for the development of large-vessel associated ischemic stroke. *J. Neurol.* **2008**, *255*, 649–654. [CrossRef] [PubMed]

32. Zhou, J.; Xu, L.; Huang, R.S.; Huang, Y.; Le, Y.; Jiang, D.; Yang, X.; Xu, W.; Huang, X.; Dong, C.; et al. Apolipoprotein A5 gene variants and the risk of coronary heart disease: A case-control study and meta-analysis. *Mol. Med. Rep.* **2013**, *8*, 1175–1182. [CrossRef] [PubMed]

33. Pi, Y.; Zhang, L.; Yang, Q.; Li, B.; Guo, L.; Fang, C.; Gao, C.; Wang, J.; Xiang, J.; Li, J. Apolipoprotein A5 gene promoter region -1131T/C polymorphism is associated with risk of ischemic stroke and elevated triglyceride levels: A meta-analysis. *Cerebrovasc. Dis.* **2012**, *33*, 558–565. [CrossRef] [PubMed]

34. Li, Y.Y.; Wu, X.Y.; Xu, J.; Qian, Y.; Zhou, C.W.; Wang, B. Apo A5 -1131T/C, FgB -455G/A, -148C/T, and CETP TaqIB gene polymorphisms and coronary artery disease in the Chinese population: A meta-analysis of 15,055 subjects. *Mol. Biol. Rep.* **2013**, *40*, 1997–2014. [CrossRef] [PubMed]

35. Sánchez-Moreno, C.; Ordovás, J.M.; Smith, C.E.; Craza, J.C.; Lee, Y.; Garaulet, M. APOA5 gene variation interacts with dietary fat intake to modulate obesity and circulating triglycerides in a Mediterranean population. *J. Nutr.* **2011**, *141*, 380–385. [CrossRef] [PubMed]

36. Zhang, M.D.; Gu, W.; Qiao, S.B.; Zhu, E.J.; Zhao, Q.M.; Ly, S.Z. Apolipoprotein E gene polymorphism and risk for coronary heart disease in the Chinese population: A meta-analysis of 61 studies including 6634 cases and 6393 controls. *PLoS ONE* **2014**, *9*, e95463. [CrossRef] [PubMed]

37. Demirkan, A.; Van Duijn, C.M.; Ugocsai, P.; Isaacs, A.; Pramstaller, P.P.; Liebisch, G.; Wilson, J.F.; Johansson, Å.; Rudan, I.; Aulchenko, Y.S.; et al. Genome-wide association study identifies novel loci associated with circulating phospho- and sphingolipid concentrations. *PLoS Genet.* **2012**, *8*, e1002490. [CrossRef] [PubMed]

38. Posadas-Sánchez, R.; Ocampo-Arcos, W.A.; López-Uribe, Á.R.; Posadas-Romero, C.; Villarreal-Molina, T.; León, E.Á.; Pérez-Hernández, N.; Rodríguez-Pérez, J.M.; Cardoso-Saldaña, G.; Medina-Urrutia, A.; et al. Hepatic lipase (LIPC) C-514T gene polymorphism is associated with cardiometabolic parameters and cardiovascular risk factors but not with fatty liver in Mexican population. *Exp. Mol. Pathol.* **2015**, *98*, 93–98. [CrossRef] [PubMed]

39. Wang, H.; Zhang, D.; Ling, J.; Lu, W.; Zhang, S.; Zhu, Y.; Lai, M. Gender specific effect of LIPC C-514T polymorphism on obesity and relationship with plasma lipid levels in Chinese children. *J. Cell. Mol. Med.* **2015**, *19*, 2296–2306. [CrossRef] [PubMed]

40. Tai, E.S.; Corella, D.; Demissie, S.; Cupples, L.A.; Coltell, O.; Schaefer, E.J.; Tucker, K.L.; Ordovas, J.M.; Framingham Heart Study. Polyunsaturated fatty acids interact with the PPARA-L162V polymorphism to affect plasma triglyceride and apolipoprotein C-III concentrations in the Framingham Heart Study. *J. Nutr.* **2005**, *135*, 397–403. [CrossRef] [PubMed]

41. Ordovas, J.M.; Corella, D.; Cupples, L.A.; Demissie, S.; Kelleher, A.; Coltell, O.; Wilson, P.W.; Schaefer, E.J.; Tucker, K. Polyunsaturated fatty acids modulate the effects of the APOA1 G-A polymorphism on HDL-cholesterol concentrations in a sex-specific manner: The Framingham Study. *Am. J. Clin. Nutr.* **2002**, *75*, 38–46. [CrossRef] [PubMed]

42. Lai, C.Q.; Corella, D.; Demissie, S.; Cupples, L.A.; Adiconis, Y.; Zhu, Y.; Parnell, L.D.; Tucker, K.L.; Ordovas, J.M. Dietary intake of n-6 fatty acids modulates effect of apolipoprotein A5 gene on plasma fasting triglycerides, remnant lipoprotein concentrations, and lipoprotein particle size: The Framingham Heart Study. *Circulation* **2006**, *113*, 2062–2070. [CrossRef] [PubMed]

 *diseases*

Review

Judicious Use of Lipid Lowering Agents in the Management of NAFLD

Umair Iqbal [1], Brandon J. Perumpail [2], Nimy John [3], Sandy Sallam [3], Neha D. Shah [3], Waiyee Kwong [3], George Cholankeril [3], Donghee Kim [3] and Aijaz Ahmed [3,*]

[1]   Department of Medicine, Geisinger Medical Center, Danville, PA 17821, USA; umairiqbal_dmc@hotmail.com
[2]   Drexel University College of Medicine, Philadelphia, PA 19104, USA; brandonperumpail@gmail.com
[3]   Division of Gastroenterology and Hepatology, Stanford University School of Medicine, Stanford,
     CA 94304, USA; nionnj@gmail.com (N.J.); SSallam@stanfordhealthcare.org (S.S.);
     NeShah@stanfordhealthcare.org (N.D.S.); WKwong@stanfordhealthcare.org (W.K.);
     georgetc@stanford.edu (G.C.); dhkimmd@stanford.edu (D.K.)
*   Correspondence: aijazahmed@stanford.edu; Tel.: +650-498-5691; Fax: +650-498-5692

Received: 9 August 2018; Accepted: 16 September 2018; Published: 24 September 2018

Abstract: Non-alcoholic fatty liver disease (NAFLD) is the most common cause of chronic liver disease in the Western world. NAFLD encompasses a spectrum of histological features, including steatosis, steatohepatitis with balloon degeneration, and hepatic fibrosis leading to cirrhosis. In patients with advanced liver damage, NAFLD is associated with an increased risk of hepatocellular carcinoma. Diabetes mellitus, hypertension, and dyslipidemia are components of metabolic syndrome and are commonly associated with NAFLD. Cardiovascular disease is the leading cause of mortality in patients with NAFLD. Therefore, it is important to pre-emptively identify and proactively treat conditions like hyperlipidemia in an effort to favorably modify the risk factors associated with cardiovascular events in patients with NAFLD. The management of hyperlipidemia has been shown to reduce cardiovascular mortality and improve histological damage/biochemical abnormalities associated with non-alcoholic steatohepatitis (NASH), a subset of NAFLD with advance liver damage. There are no formal guidelines available regarding the use of anti-hyperlipidemic drugs, as prospective data are lacking. The focus of this article is to discuss the utility of lipid-lowering drugs in patients with NAFLD.

Keywords: NAFLD; NASH; hepatic fibrosis; hyperlipidemia; statins

---

1. Introduction

Non-alcoholic fatty liver disease (NAFLD) is the most common cause of chronic liver disease in the developed world, and can progress to cirrhosis and hepatocellular carcinoma [1]. It is also the most common cause of elevated liver enzymes. Risk factors for NAFLD include obesity, diabetes mellitus, and hyperlipidemia. Obesity is also a significant risk factor for NAFLD, independent of hyperlipidemia, and weight loss has a beneficial effect on the prevention of hepatic fibrosis [2]. Patients with NAFLD are predisposed to cardiovascular mortality with studies demonstrating up to a twofold increase in the risk of cardiovascular disease in this population [3]. There are several hypotheses regarding the pathogenesis of NAFLD, which include insulin resistance, oxidative stress, and lipotoxicity (Figure 1). Hyperlipidemia is a significant risk factor for NAFLD and associated cardiovascular disease. The precise mechanism of action and the exact pathogenetic pathway on how hyperlipidemia increases the risk of NAFLD has not been elucidated, but may be related to an increased accumulation of lipids in the hepatocytes. Major sources of fatty acid delivery to hepatocytes include splanchnic lipolysis of visceral fat, lipogenesis, and the ingestion of fatty foods. Insulin resistance also downregulates low density lipoprotein (LDL)-receptor expression, leading to elevated levels of LDL.

Multiple studies have shown the efficacy of lipid-lowering drugs in patients with NAFLD and its subset non-alcoholic steatohepatitis (NASH). The aim of this article is to review the medical literature on the utility of different anti-hyperlipidemic drugs in patients with NAFLD.

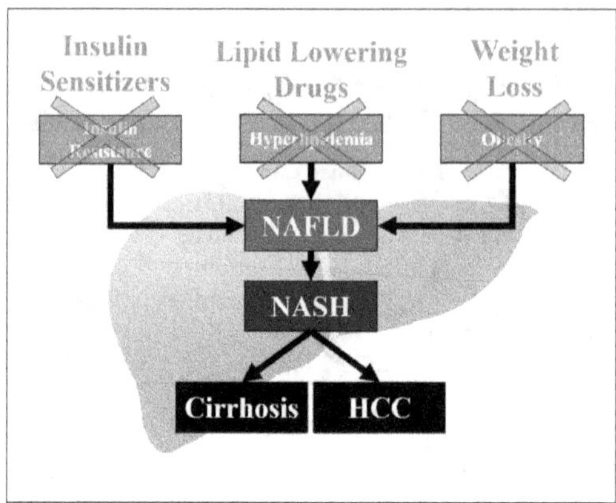

Figure 1. Pathogenesis of NAFLD/NASH and potential targets for treatment. HCC (Hepatocellular carcinoma).

2. Lipid-Lowering Drugs

2.1. Statins

Statins are the most widely used lipid lowering drugs due to their known efficacy in reducing cardiovascular mortality in patients with coronary artery disease and diabetes mellitus. These beneficial effects are not only due to the cholesterol-lowering ability of statins, but also due to their anti-inflammatory, vasodilatory (statins are nitric oxide donor), vascular remodeling, and anti-fibrotic effects independent of cholesterol-lowering activity. As inflammatory mechanisms are involved in the pathogenesis of NAFLD/NASH, statins have shown promising results decreasing liver fibrosis in patients with NAFLD. Generally, statins have been shown to be safe and effective treatment options for the indication of dyslipidemia in the context of NAFLD with the exception of Child-Pugh B and C cirrhosis and in particular if the total bilirubin level is greater than 3 mg/dL [4,5]. Table 1 summarizes the current studies evaluating the utility of different statins in NAFLD and NASH.

Table 1. Studies evaluating the utility of Statins in non-alcoholic fatty liver disease (NAFLD)/ non-alcoholic steatohepatitis (NASH).

Author's Name and Year of Publication	Drug(s) Used	Dose	Patient Population and Sample Size	Duration of Therapy	Outcome
Abel et al. [6] 2009	Simvastatin	20 mg/day	45 patients with NAFLD secondary to diabetes and metabolic syndrome	6 months	Decrease in aspartate aminotransferase (AST), and alanine aminotransferase (ALT) and low density lipoprotein (LDL) cholesterol. Liver histology was not evaluated.
Nelson et al. [7] 2009	Simvastatin	40 mg/day	16 patients with biopsy-proven NASH	12 months	No significant difference from the improvement of transaminases or hepatic fibrosis. A 26% reduction in LDL levels was seen in the simvastatin group.
Athyros et al. [8] 2006	Atorvastatin and/or Fenofibrate	20 mg/day alone or in combination with fenofibrate 200 mg/day	186 non-diabetic patients with NAFLD and metabolic syndrome	54 weeks	67% of patients on atorvastatin, 42% on fenofibrate, and 70% on combination treatment no longer had biochemical and ultrasonographic evidence of NAFLD.
Foster et al. [9] 2011	Atorvastatin	20 mg/day along with vitamin C 1g, and vitamin E 1000 IU	1005 healthy men and women between ages 50–70	Mean follow-up 3.6 years	Reduced odds of having hepatic steatosis in NAFLD patients by 71%.
Gómez-Domínguez et al. [10] 2006	Atorvastatin	10–80 mg/day	25 patients with NAFLD (only 22 completed the study)	6 months	Significant improvement in aminotransferase activities and cholesterol levels.
Kimura et al. [11] 2010	Atorvastatin	10 mg/day	43 NASH patients with dyslipidemia	12 months	Improvement in metabolic parameters related to NASH.
Kiyci et al. [12] 2003	Atorvastatin	10 mg/day	44 patients with NASH	6 months	Decrease in aminotransferase activity and fatty infiltration of the liver.
Rallidis et al. [13] 2004	Pravastatin	20 mg/day	5 patients with steatohepatitis and liver enzyme abnormalities	6 months	Improvement in histologic findings of steatohepatitis.
Hyogo et al. [14] 2011	Pitavastatin	2 mg/day	20 NASH patients with dyslipidemia	12 months	Decrease in severity of hepatic steatosis. Fibrosis stage improved in 42% of the patients.
Mihaila et al. [15] 2009	Lovastatin	10 mg/day	87 patients with NASH and dyslipidemia	4 months	Decrease in aminotransferase and cholesterol levels.
Kargiotis et al. [16] 2014	Rosuvastatin	10 mg/day	6 non-diabetic patients with NASH, metabolic syndrome, and dyslipidemia	12 months	Resolution of ultrasonographic findings of NASH. Improvement in aminotransferases; ALT and AST levels were reduced by 76% and 61%, respectively.
Kargiotis et al. [17] 2015	Rosuvastatin	10 mg/day	20 patients with NASH, metabolic syndrome, and dyslipidemia	12 months	Resolution of ultrasonographic findings of NASH.
Nakahara et al. [18] 2012	Rosuvastatin	2.5 mg/day	19 patients with NASH and dyslipidemia	24 months	Improvement of NASH-related parameters, with 33.3% patients showing improvement in non-alcoholic fatty liver disease activity score and fibrotic stage.

2.2. Simvastatin

Multiple studies have shown safety and efficacy results of simvastatin in patients with NAFLD and NASH. Abel et al. conducted a study on a small group of NAFLD patients and revealed that simvastatin at 20 mg/day for 6 months significantly decreased aspartate aminotransferase (AST), alanine aminotransferase (ALT), and LDL cholesterol levels in these patients [6]. Liver histological evaluation was not done in this study. In another randomized controlled trial, simvastatin at 40 mg/day for 12 months was compared with a placebo, which significantly reduced LDL cholesterol levels [7]. Although patient cholesterol levels were lowered in the simvastatin group, no significant differences were found in levels of transaminases or hepatic fibrosis.

2.3. Atorvastatin

Several clinical studies reported the safety and efficacy of atorvastatin in NAFLD/NASH patients [8–12]. A randomized controlled trial comparing atorvastatin with the combination of atorvastatin and fenofibrate or fenofibrate alone revealed that atorvastatin alone or in combination with fenofibrate could improve liver enzymes and ultrasonographic features of NAFLD [8]. In a St. Francis Heart study, atorvastatin along with vitamins C and E was associated with reduced likelihood of developing hepatic steatosis in patients with NAFLD [9]. Although the results of this study were promising, they may be confounded by the use of vitamin E, as vitamin E alone has been shown to be beneficial in NAFLD patients [19]. In a study by Gomez-Dominguez, 22 hyperlipidemic NAFLD patients received 10–80 mg/day of atorvastatin for 6 months and showed significant improvement in aminotransferase and cholesterol levels [10]. Kimura et al. reported the role of advanced glycation end products (AGEs) in NASH, as the levels of these products were significantly high in NASH patients. Treatment of patients with atorvastatin resulted in a significant decrease in AGE levels and an improvement in metabolic parameters related to NASH [11]. In a study done on 27 biopsy-proven NASH patients with hyperlipidemia, Kiyci et al. revealed the promising effects of atorvastatin 10 mg/day for 6 months [12]. Atorvastatin was shown to decrease aminotransferase activity and significantly decrease fatty infiltration of the liver. These studies favor the beneficial effects of atorvastatin in NAFLD patients.

2.4. Pravastatin

The current evidence regarding the use of pravastatin in NAFLD patients is lacking. In a study on five patients with biopsy-proven steatohepatitis, Rallidis et al. revealed that the use of pravastatin resulted in an improvement of hepatic histological findings [13]. Further studies—especially randomized controlled trials—are needed to further investigate the utility of pravastatin in NAFLD patients.

2.5. Pitavastatin

Pitavastatin has been indicated to be beneficial in inhibiting hepatic fibrosis in NASH rat models, but research on its utility in patients with NAFLD is still too premature to make any conclusions. Hyogo et al. conducted an open-label pilot study that revealed 2 mg/day of pitavastatin for 12 months decreased the severity of hepatic steatosis and NASH-related parameters [14]. Although the study results are promising, further studies are necessary to support these findings.

2.6. Lovastatin

In a multicentric prospective study, 10 mg/day of lovastatin for four months was shown to significantly decrease transaminases, cholesterol levels, and aminotransferase to platelet ratio indices [15]. Again, more studies are needed to support the beneficial effects of lovastatin in NAFLD patients.

2.7. Rosuvastatin

In a preliminary report on six patients with NASH, 10 mg/day of rosuvastatin for 12 months not only showed improvement in AST and ALT levels, but also the complete resolution of ultrasonographic findings of NASH in five patients [16]. Similar beneficial effects were seen in a prospective study on 20 patients, in which 10 mg/day of rosuvastatin for 12 months resulted in the complete resolution of ultrasonographic findings of NASH in 19 patients [17]. Nakahara et al. also revealed beneficial effects of 2.5 mg/day of rosuvastatin for 24 months, and showed an improvement of NASH-related parameters and histological features in some patients [18].

2.8. Non-Statin Lipid Lowering Drugs

Table 2 summarizes the current evidence regarding the utility of lipid-lowering drugs other than statins in patients with NAFLD and NASH.

Table 2. Studies evaluating the utility of lipid-lowering drugs in NAFLD/NASH.

Author's Name and Year of Publication	Drug(s) Used	Dose	Patient Population and Sample Size	Duration of Therapy	Outcome
Athyros et al. [8] 2006	Fenofibrate alone or in combination with atorvastatin	200 mg/day alone or in combination with atorvastatin 20 mg/day	186 non-diabetic patients with NAFLD and metabolic syndrome	54 weeks	42% of the patients in the fenofibrate-only treatment group had a complete resolution of biochemical and ultrasonographic evidence of NAFLD.
Basarangolu et al. [20] 1999	Gemfibrozil	600 mg/day	46 patients with NASH and liver enzyme abnormalities	4 weeks	Lower levels of ALT, AST, and gamma glutamyl transferase (GGT). No change in liver histology.
Laurir et al. [21] 1996	Clofibrate	2 g/day	40 patients with NASH	12 months	No improvement in aminotransferase activities.
Fernancez-Miranda et al. [22] 2008	Fenofibrate	200 mg/day	16 NAFLD patients	48 weeks	Decrease in insulin resistance, aminotransferases levels, and signs of metabolic syndrome. No significant effect on liver histology.
Ratzin et al. [23] 2016	Elafibranor	120 mg/day	276 patients with NASH without cirrhosis	52 weeks	Resolution of NASH without fibrosis worsening in the greater proportion of patients compared to placebo (19% vs. 12%; odds ratio = 2.31; p = 0.045).
Fabbrini et al. [24] 2010	Niacin	2000 mg/day	27 obese patients with NAFLD	16 weeks	Lowered plasma levels of very-low density lipoprotein (VLDL)-TGs but no change in intrahepatic triglycerides (TGs) content.
Yoneda et al. [25] 2010	Ezetimibe	10 mg/day	10 patients with NASH and dyslipidemia	6 months	Improvement in hepatic steatosis and histological findings. 6 out of 10 patients showed improvement in their fibrosis stage.
Park et al. [26] 2011	Ezetimibe	10 mg/day	45 NAFLD patients	24 months	Improvement in biochemical and histological abnormalities of NAFLD.
Chan et al. [27] 2010	Ezetimibe	10 mg/day	25 patients with central obesity	16 weeks	Improvement in hepatic steatosis.
Spacro et al. [28] 2008	Omega-3 (n-3) polyunsaturated fatty acids (PUFAs)	2 g/day	40 patients with NAFLD	6 months	Improvement in serum AST, ALT, triglyceride levels, and fatty liver.
Capanni et al. [29] 2006	n-3 PUFAs	1 g/day	56 patients with NAFLD	12 months	Improvement of AST, ALT, GGT, triglycerides, and ultrasonographic features of hepatic steatosis.

2.9. Fibrates

Fibrates or fibric acid derivatives are used to treat hypertriglyceridemia and primary hypercholesterolemia, mainly by activating peroxisome proliferator-activated receptor alpha (PPAR-alpha). Multiple studies have been done to evaluate the safety and efficacy of these agents in treating hyperlipidemia in NAFLD/NASH patients [8,20–22]. In a prospective open-label randomized study, Athyros et al. included 186 patients that were randomly assigned to receive either 20 mg/day of atorvastatin, 200 mg/day of fenofibrate, or both [8]. A complete resolution of biochemical and ultrasonographic evidence of NAFLD was seen in 42% of the patients in the fenofibrate group compared to 67% of patients taking atorvastatin alone, and 70% in the combination therapy group [8]. In a randomized controlled trial (RCT), Basarangolu et al. included 23 patients taking 600 mg/day of gemfibrozil for four weeks, and compared these with 23 placebo patients who were not treated with any lipid-lowering drugs. Patients in the gemfibrozil group had significantly lower levels of AST, ALT, and GGT without any significant changes in liver histology [20]. Laurin et al. evaluated the utility of the fibric acid derivative clofibrate in 16 NASH patients with hyperlipidemia taking 2 g /day [21]. The study did not find any improvement in aminotransferases, GGT, or bilirubin levels. In a pilot study on 16 biopsy-proven NAFLD patients, Fernandez-Miranda et al. revealed that 200 mg/day of fenofibrate for 48 weeks resulted in decreased insulin resistance, aminotransferases levels, and signs of metabolic syndrome [22]. Although the drug showed promising results in decreasing the proportion of patients with metabolic syndrome, its effect on liver histology was minimal. Hence, considering current evidence, we conclude that the utility of fibrates and fibric acid derivatives in ameliorating the histological features of NAFLD is still unclear and requires further studies.

2.10. Elafibranor

Elafibranor is an activator of both Peroxisome proliferator-activated receptor (PPAR)-α and PPAR-δ that has anti-inflammatory effects and helps to improve lipid metabolism and insulin resistance. In a randomized controlled double-blind trial, 276 patients with NASH were divided into three groups to receive either 80 mg/day of elafibranor, 120 mg/day of elafibranor, or a placebo for 52 weeks [23]. Aminotransferase activity, glucose levels, and inflammatory markers were significantly lower in the elafibranor 120 mg group compared to the placebo group. A post-hoc analysis of the study also revealed that elafibranor (120 mg/d for 1 year) resolved NASH without fibrosis worsening in the greater proportion of patients compared to placebo, but there was no difference in the outcome in the intention-to-treat analysis [23]. These results are encouraging for the use of elafibranor in NASH patients, but require further studies—especially randomized controlled trials—to further strengthen the evidence of these beneficial effects.

2.11. Niacin

Niacin (nicotinic acid or vitamin B-3) is used to treat vitamin deficiency (pellagra) and has lipid-lowering effects. The mechanism by which these drugs exert their lipid-lowering effects is still unclear. Fabbrini et al. reported that the combination of 200 mg/day of fenofibrate for 8 weeks and extended-release niacin at 2000 mg/day for 16 weeks lowered the plasma levels of VLDL-triglycerides, but did not alter intrahepatic triglyceride content [24]. The safety and efficacy of niacin in NAFLD patients has yet to be established, and requires further clinical studies.

2.12. Ezetimibe

Ezetimibe exerts its lipid-lowering effects by inhibiting small intestinal absorption of cholesterol via its effects on the sterol transporter Niemann–Pick C1-Like1 (NPC1L1). These molecules are also expressed in the liver, and play a role in hepatic cholesterol accumulation. The combination of ezetimibe and statins has been shown to improve LDL cholesterol levels in patients with hypercholesterolemia. In one study, Yoneda et al. reported 10 mg/day of ezetimibe for 6 months in NASH patients with hyperlipidemia resulted in significant improvement in histological findings, NAFLD activity scores, and steatosis in the liver [25]. They also revealed improvement in AST, ALT, GGT, LDL cholesterol levels, and C-reactive protein (CRP) in these patients. Similar results were seen in a study done by Park et al. regarding 45 patients with biopsy-proven NAFLD [26]. They reported that 10 mg/day of ezetimibe for 24 months improved the biochemical and histological abnormalities of NAFLD. Chan et al. also reported that weight loss along with ezetimibe was associated with an improvement in hepatic steatosis [27]. These study results favor the use of ezetimibe in NAFLD/NASH patients with hyperlipidemia, as it might ameliorate the biochemical and histological features of NAFLD.

2.13. Omega-3 (n-3) Polyunsaturated Fatty Acids (PUFAs)

The use of n-3 PUFAs increases the levels of adiponectin in the blood and decreases serum levels of triglycerides, leptin, and insulin, which promote weight loss and improve insulin resistance [28,29]. In a study of 40 patients with NAFLD, Spadro et al. reported improvement in serum AST, ALT, triglyceride levels, and fatty liver with the use of n-3 PUFAs [28]. Similar beneficial effects were seen in a study by Capanni et al., which reported that n-3 PUFA supplementation of 1 g/day for 12 months resulted in the improvement of AST, ALT, GGT, triglycerides, and ultrasonographic features of hepatic steatosis [29]. Current evidence is therefore encouraging, but further studies are needed to define the utility of n-3 PUFAs for management of NAFLD.

3. Conclusions

Statins have been shown to be most beneficial in the prevention of hepatic fibrosis in patients with NAFLD/NASH. Hence, in light of current evidence, we recommend considering statins and vitamin E along with weight loss and exercise in these patients. Although evidence regarding the utility of lipid-lowering drugs in patients with NAFLD/NASH is convincing, with most studies reporting that lipid-lowering drug use showed an improvement in hepatic steatosis, formal guidelines are lacking in this regard—mainly due to the absence of larger randomized controlled trials. Despite the need for more randomized controlled trials regarding these drugs, our review will help guide clinicians in prescribing lipid-lowering agents in NAFLD patients. This, in turn, may help decrease the morbidity and mortality associated with this serious disease.

Author Contributions: Writing-Original Draft Preparation, U.I., B.J.P., G.C., N.J., S.S., N.D.S., W.K., A.A.; Writing-Review & Editing, D.K., A.A.; Supervision, A.A.

Funding: There were no funding sources for this study.

References

1. Loomba, R.; Sanyal, A.J. The global NAFLD epidemic. *Nat. Rev. Gastroenterol. Hepatol.* **2013**, *10*, 686–690. [CrossRef] [PubMed]
2. Li, L.; Liu, D.-W.; Yan, H.-Y.; Wang, Z.-Y.; Zhao, S.-H.; Wang, B. Obesity is an independent risk factor for non-alcoholic fatty liver disease: Evidence from a meta-analysis of 21 cohort studies. *Obes. Rev.* **2016**, *17*, 510–519. [CrossRef] [PubMed]
3. Ballestri, S.; Lonardo, A.; Bonapace, S.; Byrne, C.D.; Loria, P.; Targher, G. Risk of cardiovascular, cardiac and arrhythmic complications in patients with non-alcoholic fatty liver disease. *World J. Gastroenterol.* **2014**, *20*, 1724–1745. [CrossRef] [PubMed]
4. Sigler, M.A.; Congdon, L.; Edwards, K.L. An evidence-based review of statin use in patients with nonalcoholic fatty liver disease. *Clin. Med. Insights Gastroenterol.* **2018**, *11*. [CrossRef] [PubMed]
5. Magan-Fernandez, A.; Rizzo, M.; Montalto, G.; Marchesini, G. Statins in liver disease: Not only prevention of cardiovascular events. *Expert Rev. Gastroenterol. Hepatol.* **2018**, *12*, 743–744. [CrossRef] [PubMed]

6. Abel, T.; Fehér, J.; Dinya, E.; Eldin, M.G.; Kovács, A. Safety and efficacy of combined ezetimibe/simvastatin treatment and simvastatin monotherapy in patients with non-alcoholic fatty liver disease. *Med. Sci. Monitor* **2009**, *15*, MS6–11.

7. Nelson, A.; Torres, D.M.; Morgan, A.E.; Fincke, C.; Harrison, S.A. A pilot study using simvastatin in the treatment of nonalcoholic steatohepatitis: A randomized placebo-controlled trial. *J. Clin. Gastroenterol.* **2009**, *43*, 990–994. [CrossRef] [PubMed]

8. Athyros, V.G.; Mikhailidis, D.P.; Didangelos, T.P.; Giouleme, O.I.; Liberopoulos, E.N.; Karagiannis, A.; Kakafika, A.I.; Tziomalos, K.; Burroughs, A.K.; Elisaf, M.S. Effect of multifactorial treatment on non-alcoholic fatty liver disease in metabolic syndrome: A randomised study. *Curr. Med. Res. Opin.* **2006**, *22*, 873–883. [CrossRef] [PubMed]

9. Foster, T.; Budoff, M.J.; Saab, S.; Ahmadi, N.; Gordon, C.; Guerci, A.D. Atorvastatin and antioxidants for the treatment of nonalcoholic fatty liver disease: The St Francis heart study randomized clinical trial. *Am. J. Gastroenterol.* **2011**, *106*, 71–77. [CrossRef] [PubMed]

10. Gómez-Domínguez, E.; Gisbert, J.P.; Moreno-Monteagudo, J.A.; García-Buey, L.; Moreno-Otero, R. A pilot study of atorvastatin treatment in dyslipemid, non-alcoholic fatty liver patients. *Aliment. Pharm. Therap.* **2006**, *23*, 1643–1647. [CrossRef] [PubMed]

11. Kimura, Y.; Hyogo, H.; Yamagishi, S.-I.; Takeuchi, M.; Ishitobi, T.; Nabeshima, Y.; Arihiro, K.; Chayama, K. Atorvastatin decreases serum levels of advanced glycation endproducts (AGEs) in nonalcoholic steatohepatitis (NASH) patients with dyslipidemia: Clinical usefulness of AGEs as a biomarker for the attenuation of NASH. *J. Gastroenterol.* **2010**, *45*, 750–757. [CrossRef] [PubMed]

12. Kiyici, M.; Gulten, M.; Gurel, S.; Nak, S.G.; Dolar, E.; Savci, G.; Adim, S.B.; Yerci, O.; Memik, F. Ursodeoxycholic acid and atorvastatin in the treatment of nonalcoholic steatohepatitis. *Can. J. Gastroenterol.* **2003**, *17*, 713–718. [CrossRef] [PubMed]

13. Rallidis, L.S.; Drakoulis, C.K.; Parasi, A.S. Pravastatin in patients with nonalcoholic steatohepatitis: Results of a pilot study. *Atherosclerosis* **2004**, *174*, 193–196. [CrossRef] [PubMed]

14. Hyogo, H.; Ikegami, T.; Tokushige, K.; Hashimoto, E.; Inui, K.; Matsuzaki, Y.; Tokumo, H.; Hino, F.; Tazuma, S. Efficacy of pitavastatin for the treatment of non-alcoholic steatohepatitis with dyslipidemia: An open-label, pilot study. *Hepatol. Res.* **2011**, *41*, 1057–1065. [CrossRef] [PubMed]

15. Mihaila, R.-G.; Nedelcu, L.; Fratila, O.; Rezi, E.-C.; Domnariu, C.; Deac, M. Effects of lovastatin and pentoxyphyllin in nonalcoholic steatohepatitis. *Hepato-Gastroenterol.* **2009**, *56*, 1117–1121.

16. Kargiotis, K.; Niki, K.; Athyros, V.G.; Giouleme, O.; Patsiaoura, K.; Katsiki, E.; Mikhailidis, D.P.; Karagiannis, A. Effect of rosuvastatin on non-alcoholic steatohepatitis in patients with metabolic syndrome and hypercholesterolaemia: A preliminary report. *Curr. Vasc. Pharm.* **2014**, *12*, 505–511. [CrossRef]

17. Kargiotis, K.; Athyros, V.G.; Giouleme, O.; Katsiki, N.; Katsiki, E.; Anagnostis, P.; Boutari, C.; Doumas, M.; Karagiannis, A.; Mikhailidis, D.P. Resolution of non-alcoholic steatohepatitis by rosuvastatin monotherapy in patients with metabolic syndrome. *World J. Gastroenterol.* **2015**, *21*, 7860–7868. [CrossRef] [PubMed]

18. Nakahara, T.; Hyogo, H.; Kimura, Y.; Ishitobi, T.; Arihiro, K.; Aikata, H.; Takahashi, S.; Chayama, K. Efficacy of rosuvastatin for the treatment of non-alcoholic steatohepatitis with dyslipidemia: An open-label, pilot study. *Hepatol. Res.* **2012**, *42*, 1065–1072. [CrossRef] [PubMed]

19. Sanyal, A.J.; Chalasani, N.; Kowdley, K.V.; McCullough, A.; Diehl, A.M.; Bass, N.M.; Neuschwander-Tetri, B.A. Pioglitazone, vitamin E, or placebo for nonalcoholic steatohepatitis. *New Engl. J. Med.* **2010**, *362*, 1675–1685. [CrossRef] [PubMed]

20. Basaranoglu, M.; Acbay, O.; Sonsuz, A. A controlled trial of gemfibrozil in the treatment of patients with nonalcoholic steatohepatitis. *J. Hepatol.* **1999**, *31*, 384. [CrossRef]

21. Laurin, J.; Lindor, K.D.; Crippin, J.S.; Gossard, A.; Gores, G.J.; Ludwig, J.; Rakela, J.; McGill, D.B. Ursodeoxycholic acid or clofibrate in the treatment of non-alcohol-induced steatohepatitis: A pilot study. *Hepatology* **1996**, *23*, 1464–1467. [CrossRef] [PubMed]

22. Fernández-Miranda, C.; Pérez-Carreras, M.; Colina, F.; López-Alonso, G.; Vargas, C.; Solís-Herruzo, J.A. A pilot trial of fenofibrate for the treatment of non-alcoholic fatty liver disease. *Dig. Liver Dis.* **2008**, *40*, 200–205. [CrossRef] [PubMed]

23. Ratziu, V.; Harrison, S.A.; Francque, S.; Bedossa, P.; Lehert, P.; Serfaty, L.; Romero-Gomez, M.; Boursier, J.; Abdelmalek, M.; Caldwell, S.; et al. Elafibranor, an agonist of the peroxisome proliferator-activated receptor-α and -δ, induces resolution of nonalcoholic steatohepatitis without fibrosis worsening. *Gastroenterology* **2016**, *150*, 1147–1159. [CrossRef] [PubMed]

24. Fabbrini, E.; Mohammed, B.S.; Korenblat, K.M.; Magkos, F.; McCrea, J.; Patterson, B.W.; Klein, S. Effect of fenofibrate and niacin on intrahepatic triglyceride content, very low-density lipoprotein kinetics, and insulin action in obese subjects with nonalcoholic fatty liver disease. *J. Clin. Endocrinol. Metab.* **2010**, *95*, 2727–2735. [CrossRef] [PubMed]

25. Yoneda, M.; Fujita, K.; Nozaki, Y.; Endo, H.; Takahashi, H.; Hosono, K.; Suzuki, K. Efficacy of ezetimibe for the treatment of non-alcoholic steatohepatitis: An open-label, pilot study. *Hepatol. Res.* **2010**, *40*, 566–573. [CrossRef] [PubMed]

26. Park, H.; Shima, T.; Yamaguchi, K.; Mitsuyoshi, H.; Minami, M.; Yasui, K.; Itoh, Y.; Yoshikawa, T.; Fukui, M. Efficacy of long-term ezetimibe therapy in patients with nonalcoholic fatty liver disease. *J. Gastroenterol.* **2011**, *46*, 101–107. [CrossRef] [PubMed]

27. Chan, D.C.; Watts, G.F.; Gan, S.K.; Ooi, E.M.M.; Barrett, P.H.R. Effect of ezetimibe on hepatic fat, inflammatory markers, and apolipoprotein B-100 kinetics in insulin-resistant obese subjects on a weight loss diet. *Diabetes Care* **2010**, *33*, 1134–1139. [CrossRef] [PubMed]

28. Spadaro, L.; Magliocco, O.; Spampinato, D.; Piro, S.; Oliveri, C.; Alagona, C.; Papa, G.; Rabuazzo, A.M.; Purrello, F. Effects of *N*-3 polyunsaturated fatty acids in subjects with nonalcoholic fatty liver disease. *Dig. Liver Dis.* **2008**, *40*, 194–199. [CrossRef] [PubMed]

29. Capanni, M.; Calella, F.; Biagini, M.R.; Genise, S.; Raimondi, L.; Bedogni, G.; Svegliati-Baroni, G.; Sofi, F.; Milani, S.; Abbate, R.; et al. Prolonged *N*-3 polyunsaturated fatty acid supplementation ameliorates hepatic steatosis in patients with non-alcoholic fatty liver disease: A pilot study. *Aliment. Pharm. Therap.* **2006**, *23*, 1143–1151. [CrossRef] [PubMed]

 *diseases*

Review

Inclisiran: A New Promising Agent in the Management of Hypercholesterolemia

Constantine E. Kosmas [1,*]**, Alba Muñoz Estrella** [2]**, Andreas Sourlas** [3]**, Delia Silverio** [4]**, Elizabeth Hilario** [4]**, Peter D. Montan** [4] **and Eliscer Guzman** [5]

[1]   Department of Medicine, Division of Cardiology, Mount Sinai Hospital, New York, NY 11357, USA
[2]   Department of Medicine, Mount Sinai St. Luke's-West Hospital, New York, NY 10025, USA; alba.munoz.estrella@gmail.com
[3]   School of Medicine, University of Crete, 71003 Heraklion, Greece; med3553@edu.med.uoc.gr
[4]   Cardiology Clinic, Cardiology Unlimited, PC, New York, NY 10033, USA; deliasg10@gmail.com (D.S.); bethhilario@hotmail.com (E.H.); drpetermontan@gmail.com (P.D.M.)
[5]   Department of Medicine, Division of Cardiology, Montefiore Medical Center, Bronx, NY 10467, USA; eliscer@hotmail.com
*   Correspondence: cekosmas1@gmail.com; Tel.: +1-646-734-7969

Academic Editor: Ahmed Bakillah
Received: 29 June 2018; Accepted: 11 July 2018; Published: 13 July 2018

Abstract: The discovery of proprotein convertase subtilisin-kexin type 9 (PCSK9), a serine protease which binds to the low-density lipoprotein (LDL) receptors and targets the receptors for lysosomal degradation, offered an additional route through which plasma LDL-cholesterol (LDL-C) levels can be controlled. Initially, the therapeutic approaches to reduce circulating levels of PCSK9 were focused on the use of monoclonal antibodies. To that effect, evolocumab and alirocumab, two human monoclonal antibodies directed against PCSK9, given on a background of statin therapy, have been shown to markedly decrease LDL-C levels and significantly reduce cardiovascular risk. The small interfering RNA (siRNA) molecules have been used recently to target the hepatic production of PCSK9. siRNA interferes with the expression of specific genes with complementary nucleotide sequences by affecting the degradation of mRNA post-transcription, thus preventing translation. Inclisiran is a long-acting, synthetic siRNA directed against PCSK9 and it has been shown to significantly decrease hepatic production of PCSK9 and cause a marked reduction in LDL-C levels. This review aims to present and discuss the current clinical and scientific evidence pertaining to inclisiran, which is a new promising agent in the management of hypercholesterolemia.

Keywords: proprotein convertase subtilisin-kexin type 9 (PCSK9); small interfering RNA (siRNA); inclisiran; low-density lipoprotein-cholesterol (LDL-C); cardiovascular disease (CVD); cardiovascular risk

---

1. Introduction

Cardiovascular disease (CVD) is the number one cause of death worldwide. An estimated 17.7 million people died from CVD in 2015, representing 31% of all global deaths [1]. Hypercholesterolemia is a major known risk factor for cardiovascular disease and low-density lipoprotein cholesterol (LDL-C) lowering has been unequivocally shown to cause a significant reduction in cardiovascular risk in both primary and secondary prevention [2,3]. Statins are the standard of care and have a proven efficacy in LDL-C lowering and in the reduction of CVD risk [4,5]. However, there is considerable variability in individual responses to statins [6] and many individuals at risk for CVD fail to achieve LDL-C goals [7,8]. Furthermore, several patients demonstrate intolerance to statins, mostly due to myalgias and weakness [9]. These factors necessitate research for the development of new additional therapies with a favorable side effect profile that would improve our ability to achieve LDL-C goals and decrease CVD risk [10].

The discovery of proprotein convertase subtilisin-kexin type 9 (PCSK9) in 2003 [11], a serine protease which binds to the LDL receptors and targets the receptors for lysosomal degradation, thereby reducing their recycling and decreasing the removal rate of circulating LDL-C [12], offered an additional route through which plasma LDL-C levels can be controlled [13]. Initially, the therapeutic approaches to reduce circulating levels of PCSK9 were focused on the use of monoclonal antibodies. To that effect, evolocumab and alirocumab, two human monoclonal antibodies directed against PCSK9, given on a background of statin therapy, have been shown to markedly decrease LDL-C levels and significantly reduce cardiovascular risk [14,15]. Circulating PCSK9 is generated mainly by the liver; hence, therapeutic agents curtailing the hepatic production of PCSK9 may provide an alternative to the use of monoclonal antibodies. Inclisiran is a chemically synthesized small interfering RNA (siRNA) molecule, which targets the hepatic production of PCSK9 and is currently under investigation for its LDL-C lowering effect and potential for cardiovascular risk reduction.

2. siRNAs: Mechanism of Action

siRNAs are ~20–30 nucleotide RNA molecules, which lately have emerged as critical regulators in the expression and function of eukaryotic genomes. These molecules, which may be active in both the somatic and germline lineages of different eukaryotic species, are involved in the regulation of endogenous genes and in the protection of the genome from invasive nucleic acids [16]. In most mammalian cells, long double-stranded RNA induces an interferon response, which contributes to the antiviral defense. This interferon response prompts a generalized shutdown of protein synthesis. As a result, long double-stranded RNA cannot be used for specific gene silencing. On the contrary, siRNAs can evade the radar of the mammalian interferon response and produce strong and specific gene silencing [17]. More specifically, siRNA interferes with the expression of specific genes with complementary nucleotide sequences by affecting the degradation of mRNA post-transcription, thus preventing translation [18].

3. Inclisiran

The siRNA molecule has been used recently to decrease PCSK9 levels. The siRNA molecules follow the natural pathway of RNA interference (RNAi) by binding intracellularly to the RNA-induced silencing complex (RISC), thus enabling it to cleave messenger RNA (mRNA) molecules specifically encoding PCSK9 [19].

Inclisiran is a long-acting, synthetic siRNA directed against PCSK9, which is conjugated to triantennary *N*-acetylgalactosamine carbohydrates. These carbohydrates bind to abundant liver-expressed asialoglycoprotein receptors, leading to the uptake of inclisiran specifically into the hepatocytes [20,21].

A schematic of the mechanism of action of inclisiran (along with the action of PCSK9, the hepatic production of which is significantly decreased by inclisiran) is shown in Figure 1.

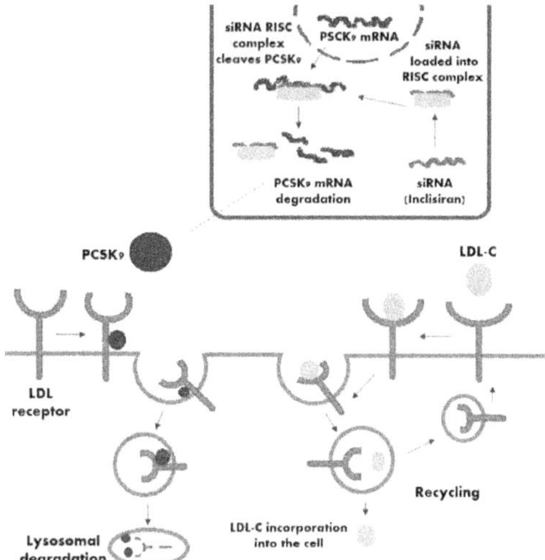

Figure 1. Mechanism of action of inclisiran in conjunction with the action of PCSK9.

4. Clinical Trials with Inclisiran

In a randomized, single-blind, placebo-controlled, phase 1 dose-escalation study in healthy adult volunteers with serum LDL-C levels $\geq$ 3.00 mmol/L (116 mg/dL), participants were randomly assigned in a 3:1 ratio to receive one dose of intravenous ALN-PCS (inclisiran), with doses ranging from 0.015 to 0.400 mg/kg, versus placebo. In the group of participants who received treatment with inclisiran at the dose of 0.400 mg/kg, there was a significant mean reduction in circulating PCSK9 and LDL-C levels by 70% and 40%, respectively, as compared with placebo (p < 0.0001 for both comparisons). The proportions of patients affected by treatment-emergent adverse events were similar in the inclisiran and placebo groups (79% vs. 88%, respectively) [22].

In another phase 1 trial, several different doses of inclisiran or placebo were administered subcutaneously to healthy volunteers with an LDL-C level of at least 100 mg/dL. The doses of inclisiran tried were single-dose-injections (25, 100, 300, 500, and 800 mg) or multiple-dose injections (125 mg weekly for four doses, 250 mg every other week for two doses, and 300 or 500 mg monthly for two doses) with or without concurrent statin therapy. Single doses of inclisiran $\geq$ 300 mg decreased the PCSK9 level by up to a least-squares mean reduction of 74.5% from baseline to day 84, whereas single doses of $\geq$100 mg lowered LDL cholesterol-C by up to a least-squares mean reduction of 50.6% from baseline. Reductions in the PCSK9 and LDL-C levels were maintained at day 180 for doses of $\geq$300 mg. All multiple-dose regimens of inclisiran reduced the levels of PCSK9 and LDL-C by up to a least-squares mean reduction of 83.8% and 59.7% from baseline to day 84, respectively. There were no serious adverse events observed with inclisiran and the most common adverse events were cough, musculoskeletal pain, nasopharyngitis, headache, back pain, and diarrhea [23].

ORION-1 was a phase 2, multicenter, double-blind, placebo-controlled, multiple-ascending-dose trial of inclisiran, administered as a subcutaneous injection, in patients at high risk for CVD with LDL-C levels > 70 mg/dL in the presence of a history of atherosclerotic CVD (ASCVD) or > 100 mg/dL in the absence of a history of ASCVD. A total of 501 patients were randomly assigned to receive a single dose of placebo or 200, 300, or 500 mg of inclisiran, or two doses of placebo or 100, 200, or 300 mg of inclisiran at days 1 and 90. The primary end point was the change in LDL-C level from baseline level at 180 days. At enrollment, 73% of the patients were on statin therapy, and 31% of the patients were on treatment with ezetimibe. At day 180, the least-squares mean reductions in LDL-C levels were 27.9 to 41.9% in the patients who received a single dose of inclisiran and 35.5% to 52.6% in the patients who received two doses ($p < 0.001$ for all comparisons vs. placebo). The greatest reduction in LDL-C levels was attained with the two-dose 300-mg inclisiran regimen and 48% of the patients who received that regimen had an LDL-C level < 50 mg/dL at day 180. Furthermore, the two-dose 300-mg inclisiran regimen caused a least-squares mean reduction in PCSK9 levels by 69.1% ($p < 0.001$ vs. placebo) and decreased high sensitivity C-reactive protein (hsCRP) by 16.7% ($p < 0.05$). Serious adverse events occurred in 11% of the patients who received inclisiran and in 8% of the patients who received placebo. Injection-site reactions occurred in 4% of patients who received one dose and in 7% of patients who received two doses of inclisiran [24]. Thus, in this phase 2 trial, inclisiran produced significant reductions in LDL-C and PCSK9 levels with an acceptable side effect profile, as compared to placebo. However, given the relatively small number of patients and the short duration of the study, no definitive conclusions can be drawn regarding the long-term side effect profile of inclisiran and further larger studies with a longer follow-up period are required for that reason.

ORION-11 is an ongoing placebo-controlled, double-blind, randomized, phase 3 trial with inclisiran. The study will enroll individuals with ASCVD or ASCVD-risk equivalents and elevated LDL-C despite maximum tolerated dose of LDL-C lowering therapies in order to evaluate the efficacy, safety, and tolerability of subcutaneous inclisiran injection(s). The study will be an international multicenter study (non-United States). Currently, 1617 participants have been enrolled. The patients will be randomly assigned to either inclisiran or placebo. Doses of 300 mg of inclisiran sodium (equivalent to 284 mg of inclisiran) will be administered as subcutaneous injections on day 1, day 90, and then every 6 months. Primary outcomes include percentage change in LDL-C from baseline to day 510 and time-adjusted percent change in LDL-C levels from baseline between day 90 and day 540. The secondary outcomes include absolute change in LDL-C from baseline to day 510, time-adjusted absolute change in LDL-C from baseline between day 90 and day 540, as well as the percentage changes in the levels of PCSK9, total cholesterol, apolipoprotein B (ApoB), and non-high-density lipoprotein cholesterol (non-HDL-C) from baseline to day 510 [25].

ORION-11 trial is one of the four phase 3 pivotal trials with inclisiran. Others include the ORION-10 trial with approximately 1500 ASCVD patients in North America, the ORION-9 trial with approximately 400 patients with heterozygous familial hypercholesterolemia (FH) in North America, Europe, Israel, and South Africa, as well as the ORION-5 trial with 60 patients with homozygous FH in Europe, Middle East, and North America [26].

5. Conclusions and Future Directions

Medical interventions targeting PCSK9 have emerged as a very promising therapeutic strategy in the management of hyperlipidemia. By inhibiting the expression of PCSK9, significant reductions in the levels of LDL-C have been obtained that may lead to a reduction of cardiovascular risk [20]. As it was mentioned before, two PCSK9 inhibitors (evolocumab and alirocumab) have already been tested in large outcome trials and have been shown to significantly decrease cardiovascular risk [14,15].

On the other hand, inclisiran, a synthetic siRNA molecule, has been shown to significantly decrease hepatic production of PCSK9 and cause a marked reduction in LDL-C levels [22–24], although, up to date, there are no available studies showing the effect of inclisiran on intermediate CVD markers, such as intima-media thickness of the carotid artery (CIMT), arterial flow-mediated dilation (FMD), or arterial pulse wave velocity (PWV). Advantages of inclisiran over monoclonal antibodies directed against PCSK9 include its infrequent administration (twice a year vs. 12–26 injections per year for PCSK9 inhibitors), as well as the fact that anti-PCSK9 monoclonal antibodies act at a plasma level, whereas inclisiran acts at the intracellular level of hepatocytes to mitigate the levels of LDL-C and PCSK9 [27].

Furthermore, inclisiran appears to have a relatively benign side effect profile, as shown in the ORION-1 trial. There were only rare symptoms of immune activation, such as flu-like symptoms, which is often a concern with RNA-targeting therapies. In addition, inclisiran did not adversely affect platelet levels, in contrast to other recent reports from studies of antisense oligonucleotides and other siRNA molecules [24,28]. However, the ongoing ORION-11 trial [25] is expected to better define the long-term side effect profile of inclisiran.

Notwithstanding, it is not yet ascertained that the marked LDL-C reductions attained with inclisiran would definitely translate into a reduction in CVD risk, and thus, large outcome trials would need to be conducted in the future for that reason.

Other potential future therapeutic strategies targeting PCSK9, which are currently in the initial stages of development, include small molecule inhibitors that disrupt the processing of PCSK9, the use of adnectins, which block the binding of PCSK9 to the LDL receptor [20,29], as well as the AT04A vaccine, which is currently being tested in a phase 1 clinical trial [20,30].

Funding: This research received no external funding.

Conflicts of Interest: C.E.K. and E.G. are members of the Dyslipidemia Speaker Bureau of Amgen, Inc. A.M.E., A.S., D.S., E.H. and P.D.M. declare no conflicts of interest.

References

1. World Health Organization. Cardiovascular Diseases (CVDs). Available online: http://www.who.int/news-room/fact-sheets/detail/cardiovascular-diseases-(cvds) (accessed on 17 May 2017).
2. O'Keefe, J.H., Jr.; Cordain, L.; Harris, W.H.; Moe, R.M.; Vogel, R. Optimal low-density lipoprotein is 50 to 70 mg/dL: Lower is better and physiologically normal. *J. Am. Coll. Cardiol.* **2004**, *43*, 2142–2146. [CrossRef] [PubMed]
3. LaRosa, J.C.; Grundy, S.M.; Waters, D.D.; Shear, C.; Barter, P.; Fruchart, J.C.; Gotto, A.M.; Greten, H.; Kastelein, J.J.; Shepherd, J.; et al. Intensive lipid lowering with atorvastatin in patients with stable coronary disease. *N. Engl. J. Med.* **2005**, *352*, 1425–1435. [CrossRef] [PubMed]
4. Waters, D.D. What the statin trials have taught us. *Am. J. Cardiol.* **2006**, *98*, 129–134. [CrossRef] [PubMed]
5. Grundy, S.M.; Cleeman, J.I.; Merz, C.N.; Brewer, H.B., Jr.; Clark, L.T.; Hunninghake, D.B.; Pasternak, R.C.; Smith, S.C., Jr.; Stone, N.J.; National Heart, Lung, and Blood Institute; et al. Implications of recent clinical trials for the National Cholesterol Education Program Adult Treatment Panel III guidelines. *Circulation* **2004**, *110*, 227–239. [CrossRef] [PubMed]
6. Ridker, P.M.; Mora, S.; Rose, L.; JUPITER Trial Study Group. Percent reduction in LDL cholesterol following high-intensity statin therapy: Potential implications for guidelines and for the prescription of emerging lipid-lowering agents. *Eur. Heart J.* **2016**, *7*, 1373–1379. [CrossRef] [PubMed]
7. Martin, S.S.; Gosch, K.; Kulkarni, K.R.; Spertus, J.A.; Mathews, R.; Ho, P.M.; Maddox, T.M.; Newby, L.K.; Alexander, K.P.; Wang, T.Y. Modifiable factors associated with failure to attain low-density lipoprotein cholesterol goal at 6 months after acute myocardial infarction. *Am. Heart J.* **2013**, *165*, 26–33. [CrossRef] [PubMed]
8. Virani, S.S.; Woodard, L.D.; Chitwood, S.S.; Landrum, C.R.; Urech, T.H.; Wang, D.; Murawsky, J.; Ballantyne, C.M.; Petersen, L.A. Frequency and correlates of treatment intensification for elevated cholesterol levels in patients with cardiovascular disease. *Am. Heart J.* **2011**, *162*, 725–732. [CrossRef] [PubMed]
9. Cohen, J.D.; Brinton, E.A.; Ito, M.K.; Jacobson, T.A. Understanding Statin Use in America and Gaps in Patient Education (USAGE): An internet-based survey of 10,138 current and former statin users. *J. Clin. Lipidol.* **2012**, *6*, 208–215. [CrossRef] [PubMed]
10. Kosmas, C.E.; Frishman, W.H. New and emerging LDL cholesterol-lowering drugs. *Am. J. Ther.* **2015**, *22*, 234–241. [CrossRef] [PubMed]
11. Seidah, N.G.; Benjannet, S.; Wickham, L.; Marcinkiewicz, J.; Jasmin, S.B.; Stifani, S.; Basak, A.; Prat, A.; Chretien, M. The secretory proprotein convertase neural apoptosis-regulated convertase 1 (NARC-1): Liver regeneration and neuronal differentiation. *Proc. Natl. Acad. Sci. USA* **2003**, *100*, 928–933. [CrossRef] [PubMed]
12. Leren, T.P. Sorting an LDL receptor with bound PCSK9 to intracellular degradation. *Atherosclerosis* **2014**, *237*, 76–81. [CrossRef] [PubMed]
13. Cohen, J.C.; Boerwinkle, E.; Mosley, T.H., Jr.; Hobbs, H.H. Sequence variations in PCSK9, low LDL, and protection against coronary heart disease. *N. Engl. J. Med.* **2006**, *354*, 1264–1272. [CrossRef] [PubMed]

14. Sabatine, M.S.; Giugliano, R.P.; Keech, A.C.; Honarpour, N.; Wiviott, S.D.; Murphy, S.A.; Kuder, J.F.; Wang, H.; Liu, T.; Wasserman, S.M.; et al. Evolocumab and Clinical Outcomes in Patients with Cardiovascular Disease. *N. Engl. J. Med.* **2017**, *376*, 1713–1722. [CrossRef] [PubMed]

15. Steg, P.G. Evaluation of Cardiovascular Outcomes after an Acute Coronary Syndrome during Treatment with Alirocumab—ODYSSEY OUTCOMES. In Proceedings of the American College of Cardiology Annual Scientific Session (ACC 2018), Orlando, FL, USA, 10–12 March 2018.

16. Carthew, R.W.; Sontheimer, E.J. Origins and Mechanisms of miRNAs and siRNAs. *Cell* **2009**, *136*, 642–655. [CrossRef] [PubMed]

17. Bernards, R. Exploring the uses of RNAi-gene knockdown and the Nobel Prize. *N. Engl. J. Med.* **2006**, *355*, 2391–2393. [CrossRef] [PubMed]

18. Agrawal, N.; Dasaradhi, P.V.; Mohmmed, A.; Malhotra, P.; Bhatnagar, R.K.; Mukherjee, S.K. RNA interference: Biology, mechanism, and applications. *Microbiol. Mol. Biol. Rev.* **2003**, *67*, 657–685. [CrossRef] [PubMed]

19. Ciccarelli, G.; D'Elia, S.; De Paulis, M.; Golino, P.; Cimmino, G. Lipid Target in Very High-Risk Cardiovascular Patients: Lesson from PCSK9 Monoclonal Antibodies. *Diseases* **2018**, *6*, 22. [CrossRef] [PubMed]

20. Kosmas, C.E.; DeJesus, E.; Morcelo, R.; Garcia, F.; Montan, P.D.; Guzman, E. Lipid-lowering interventions targeting proprotein convertase subtilisin/kexin type 9 (PCSK9): An emerging chapter in lipid-lowering therapy. *Drugs Context* **2017**, *6*, 212511. [CrossRef] [PubMed]

21. Nair, J.K.; Willoughby, J.L.; Chan, A.; Charisse, K.; Alam, M.R.; Wang, Q.; Hoekstra, M.; Kandasamy, P.; Kel'in, A.V.; Milstein, S.; et al. Multivalent N-acetylgalactosamine-conjugated siRNA localizes in hepatocytes and elicits robust RNAi-mediated gene silencing. *J. Am. Chem. Soc.* **2014**, *136*, 16958–16961. [CrossRef] [PubMed]

22. Fitzgerald, K.; Frank-Kamenetsky, M.; Shulga-Morskaya, S.; Liebow, A.; Bettencourt, B.R.; Sutherland, J.E.; Hutabarat, R.M.; Clausen, V.A.; Karsten, V.; Cehelsky, J.; et al. Effect of an RNA interference drug on the synthesis of proprotein convertase subtilisin/kexin type9 (PCSK9) and the concentration of serum LDL cholesterol in healthy volunteers: A randomised, single-blind, placebo-controlled, phase 1 trial. *Lancet* **2014**, *383*, 60–68. [CrossRef]

23. Fitzgerald, K.; White, S.; Borodovsky, A.; Bettencourt, B.R.; Strahs, A.; Clausen, V.; Wijngaard, P.; Horton, J.D.; Taubel, J.; Brooks, A.; et al. A Highly Durable RNAi Therapeutic Inhibitor of PCSK9. *N. Engl. J. Med.* **2017**, *376*, 41–51. [CrossRef] [PubMed]

24. Ray, K.K.; Landmesser, U.; Leiter, L.A.; Kallend, D.; Dufour, R.; Karakas, M.; Hall, T.; Troquay, R.P.; Turner, T.; Visseren, F.L.; et al. Inclisiran in patients at high cardiovascular risk with elevated LDL cholesterol. *N. Engl. J. Med.* **2017**, *376*, 1430–1440. [CrossRef] [PubMed]

25. Inclisiran for Subjects with ACSVD or ACSVD-Risk Equivalents and Elevated Low-Density Lipoprotein Cholesterol (ORION-11). ClinicalTrials.gov Identifier: NCT03400800. Available online: http://clinicaltrials. gov/ct2/show/NCT03400800 (accessed on 26 April 2018).

26. The Medicines Company. The Medicines Company and Alnylam Pharmaceuticals Announce Initiation of Phase III Clinical Trials of Inclisiran. Available online: http://www.themedicinescompany.com/investors/ news/medicines-company-and-alnylam-pharmaceuticals-announce-initiation-phase-iii-clinical (accessed on 6 November 2017).

27. Bandyopadhyay, D.; Hajra, A.; Ashish, K.; Qureshi, A.; Ball, S. New hope for hyperlipidemia management: Inclisiran. *J. Cardiol.* **2018**, *71*, 523–524. [CrossRef] [PubMed]

28. Chi, X.; Gatti, P.; Papoian, T. Safety of antisense oligonucleotide and siRNA-based therapeutics. *Drug Discov. Today* **2017**, *22*, 823–833. [CrossRef] [PubMed]

29. Giugliano, R.P.; Sabatine, M.S. Are PCSK9 inhibitors the next breakthrough in the cardiovascular field? *J. Am. Coll. Cardiol.* **2015**, *65*, 2638–2651. [CrossRef] [PubMed]

30. Landlinger, C.; Pouwer, M.G.; Juno, C.; van der Hoorn, J.W.A.; Pieterman, E.J.; Jukema, J.W.; Staffler, G.; Princen, H.M.G.; Galabova, G. The AT04A vaccine against proprotein convertase subtilisin/kexin type 9 reduces total cholesterol, vascular inflammation, and atherosclerosis in APOE*3Leiden.CETP mice. *Eur. Heart J.* **2017**, *38*, 2499–2507. [CrossRef] [PubMed]

 *diseases*

Review

CVD Risk Stratification in the PCSK9 Era: Is There a Role for LDL Subfractions?

Christian Abendstein Kjellmo [1,*], Anders Hovland [1,2] and Knut Tore Lappegård [1,2]

1   Division of Internal Medicine, Nordland Hospital, N-8092 Bodø, Norway; anders.w.hovland@gmail.com (A.H.); knut.tore.lappegard@gmail.com (K.T.L.)

2   Department of Clinical Medicine, University of Tromsø, N-9037 Tromsø, Norway

*   Correspondence: kjellmo@gmail.com; Tel.: +47-40-875-390

Received: 4 May 2018; Accepted: 24 May 2018; Published: 27 May 2018

Abstract: Proprotein convertase subtilisin/kexin type 9 (PCSK9) inhibitors reduce the risk of cardiovascular events and all-cause mortality in patients at high risk of cardiovascular disease (CVD). Due to high costs and unknown long-term adverse effects, critical evaluation of patients considered for PCSK9 inhibitors is important. It has been proposed that measuring low-density lipoprotein (LDL) subfractions, or LDL particle numbers (LDL-P), could be of value in CVD risk assessment and may identify patients at high risk of CVD. This review evaluates the evidence for the use of LDL subfractions, or LDL-P, when assessing CVD risk in patients for whom PCSK9 inhibitors are considered as a lipid-lowering therapy. Numerous methods for measuring LDL subfractions and LDL-P are available, but several factors limit their availability. A lack of standardization makes comparison between the different methods challenging. Longitudinal population-based studies have found an independent association between different LDL subfractions, LDL-P, and an increased risk of cardiovascular events, but definitive evidence that these measurements add predictive value to the standard risk markers is lacking. No studies have proven that these measurements improve clinical outcomes. PCSK9 inhibitors seem to be effective at lowering all LDL subfractions and LDL-P, but any evidence that measuring LDL subfractions and LDL-P yield clinically useful information is lacking. Such analyses are currently not recommended when considering whether to initiate PCKS9 inhibitors in patients at risk of CVD.

Keywords: PCSK9; proprotein convertase subtilisin/kexin type 9; LDL subfractions; sdLDL; cardiovascular disease; risk stratification

---

1. Introduction

Despite major advances in the prevention and treatment of cardiovascular diseases (CVD) over the last few decades, CVD continues to be the leading global cause of death and morbidity [1]. Several different guidelines for CVD prevention are available and the recommended overall strategy is the targeting of modifiable risk factors in high risk patients [2,3].

Of the multiple modifiable risk factors associated with cardiovascular disease [4], low-density lipoprotein (LDL) is the most intensively studied and a causal relationship with the development of CVD has been established [5]. Managing LDL-related risk is emphasized in all CVD prevention guidelines by recommending lipid-lowering therapy, usually statins, to all patients for secondary prevention, and to high-risk patients for primary prevention [6]. The guidelines for CVD prevention are not unified in their recommendations on what lipoprotein measurement to use in risk assessment and as a target of therapy [2,3,7]. Non-high-density lipoprotein cholesterol (non-HDL-C) is the lipoprotein measurement recommended for risk assessment in most guidelines, as it reflects all the cholesterol mass with atherogenic potential and avoids the biases that might arise when using the Friedewald formula to calculate LDL cholesterol (LDL-C) [8]. LDL-C is still the most widely-recommended primary target of therapy. Both metrics are included in the standard lipid panel, which is readily available at most clinical laboratories. Despite its central role in CVD pathophysiology, the value of both non-HDL-C and LDL-C in CVD risk stratification is limited as a significant proportion of patients who develop CVD have levels within the "normal" range [9]. Due to this, there has been intensive research into whether different advanced lipoprotein testing methods may improve cardiovascular risk prediction.

LDL-C is a measure of the total cholesterol content in LDL particles. LDL-C and LDL particle number (LDL-P) is usually highly correlated [10]. Under certain circumstances, notably in patients with diabetes, metabolic syndrome, or hypertriglyceridemia, LDL-C and LDL-P can become discordant due to the predominance of small dense cholesterol-depleted LDL-particles (sdLDL) [11]. In these patients, LDL-C might not accurately reflect the LDL-related risk for cardiovascular disease, and studies have shown that LDL-P has a stronger association with CVD risk compared to LDL-C in patients with discordant levels of LDL-C and LDL-P [10,11]. Due to this fact, it has been proposed that measuring subfractions or the particle number of LDL, might enhance CVD risk assessment in the general population and detect residual risk in patients already receiving lipid-lowering therapy.

Recent advances in lipid lowering therapies, with the development of proprotein convertase subtilisin/kexin type 9 (PCSK9) inhibitors [12], has reignited attention on CVD-risk stratification. Clinicians now have the tools to reduce LDL to very low levels, but the costs are significant and potential side effects have only been evaluated in relatively short-term studies. For this reason, PCKS9 inhibitors are currently only recommended to patients at a very high risk, such as patients with familial hypercholesterolemia (FH), statin-intolerant patients in secondary prevention, or in secondary prevention for patients with high residual risk [13,14].

In this review, we sought to evaluate the evidence for the use of LDL subfractions in CVD risk assessment in general, and to assess if the available methods for LDL subfractioning could be of value for clinicians in the decision of whether to initiate PCKS9 therapy in patients.

2. LDL Subfractions—And How to Separate Them

LDLs are broadly defined as lipoproteins with a density in the range of 1.019–1.063 g/mL, and each particle containing one apolipoprotein B (apoB) molecule (Figure 1). LDL particles are heterogeneous with respect to size, density, and composition, and can be separated based on various physicochemical properties depending on the protein purification technique used. The methods commonly used in the published literature on LDL subfractions are usually based on gel electrophoresis (GE), nuclear magnetic resonance (NMR), ultracentrifugation (UC), or ion mobility (IM), but several other methods are also available [15,16].

The accessibility of these methods is limited. GE, NMR, UC, and IM are technically demanding, expensive, and too labor-intensive for routine use in clinical laboratories. Most methods used in published studies are patented laboratory developed tests, which are only available at one laboratory or a limited number of laboratories. Two exceptions include a simple and less-expensive tube gel electrophoresis-system (Lipoprint®) [17] and a homogeneous assay (HA) adaptable to autoanalyzers [16].

The different methods use various terms to describe LDL particles, their distribution, and characteristics, including: LDL subfractions, LDL subclasses, LDL particle and subfraction concentration, LDL particle diameter, LDL peak diameter, and others [18–20]. These terms describe overlapping attributes of the LDL particles. The potential for confusion is significant, and in this review we will use the generic term LDL subfractions. LDL particle number (LDL-P) is a measurement of the total number of LDL particles across all subfractions, but will be included in this review as NMR and IM methods report it. ApoB is another indirect measurement of LDL-P and the ratio between serum cholesterol and apoB has been proposed as a metric to determine the prevalence of sdLDL [21]. However, apoB will not be included in this review as apoB is not reported by the various methods that determine LDL subfractions.

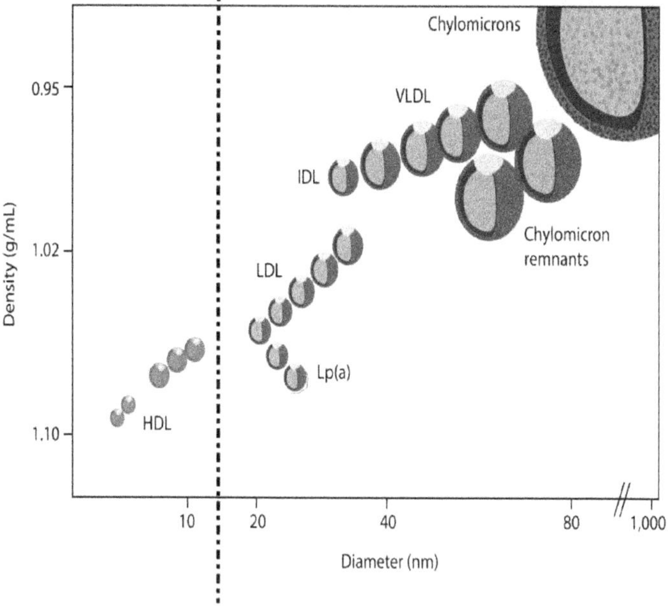

Figure 1. Relative size and density of the major plasma lipoproteins and their subfractions.

The recognition of the different LDL subfractions has led to the description of two distinct patterns of phenotypes that are reported by most LDL subfractioning methods available today [22]: Phenotype "A", with a predominance of large buoyant LDL-particles and phenotype "B", with a predominance of small, dense LDL particles, which is suggested as a more atherogenic phenotype.

3. Comparability of Methods for LDL Subfractioning

Presently, there is no method for the determination of LDL subfractions that is considered a reference or a gold standard. The different methods commonly available (NMR, UC, GE, and IM) separate LDL particles based on different characteristics that are not directly comparable; GE separates LDL particles based on size and charge, UC based on density, NMR measures methyl group signals from lipoprotein particles and calculates the LDL particle number and size, while IM separates lipoprotein particles using gas-phase electrophoresis and directly counting the size-separated particles.

In 2009, Chung et al. [23] published a comprehensive review of the comparability between several different methods for LDL subfractioning, including UC, NMR, and segmented gradient GE (sGGE). This review raised several issues. These methods separate LDL particles based on completely different physicochemical properties, and, consequently, it is not possible to determine whether the different methods are measuring the same subfractions. Different thresholds of LDL size and particle number are used to describe LDL particle distributions and LDL phenotypes, further obfuscating a reasonable comparison of results between methods. Furthermore, the wide range of agreement between the methods reported (7–94% concordance for classifying LDL phenotypes) was suggested to be due to a lack of standardization. Lastly, no study has compared the diagnostic accuracy, efficacy, or clinical value of these methods to individuals' clinical outcomes.

Chung et al. concluded that, to permit an accurate description of the similarities and differences among the different LDL subfractioning methods, a reference material, accepted as appropriate, accurate and reliable, needed to be developed. After the establishment of a consensus reference method, epidemiological studies and clinical trials are required to examine the clinical value of LDL subfractioning to predict the individual patients' clinical outcomes [23].

Since 2009, no reference material, nor a consensus reference method, has been established. IM has been introduced and has been used in the larger population-based studies [24,25], adding to the complexity of LDL subfractioning. A few studies, published since 2009, have looked at the comparability between different methods. Two studies have reported high rates of correlation, including a 98% agreement between IM and sGGE [26,27], but as a healthy study population was used the results need to be confirmed in a population relevant for CVD risk assessment. Williams et al. reported that both sGGE, NMR, and IM confirmed the association between small LDL particles and greater atherosclerotic progression, but the correlations between the methods varied significantly [28].

In contrast, international standards exist for assays measuring HDL-C, total cholesterol, LDL-C, apoB and apolipoprotein A1, and Lipoprotein(a). [29]

4. The Atherogenicity of Small, Dense LDL Particles

The first step in the initiation of atherosclerotic disease is the subendothelial retention of apoB-containing lipoproteins, predominantly LDL, in the arterial wall [30]. Elevated levels of apoB-containing lipoproteins in the blood increases the risk of atherosclerotic disease, in a dose-dependent matter [31]. One possible explanation for increased CVD risk associated with elevated levels of sdLDL or LDL-P could be that these measures more accurately reflect the number of apoB-containing lipoproteins compared to LDL-C.

There is also evidence indicating that sdLDL-particles have an increased atherogenic potential when compared to larger LDL particles [32]. The circulation time of sdLDLs is proposed to be longer, compared to larger LDL-particles, due to their impaired interaction with the LDL-receptor (LDLR) [33]. An increased susceptibility of sdLDL to undergo atherogenic modification in blood plasma, such as desialyation, glycation, and oxidation, has also been reported [34,35] and in-vitro studies have shown that sdLDL particles are more avidly taken up by macrophages, have a greater propensity for transport into the artery wall, and have a greater binding potential to proteoglycans in the artery wall [36,37].

5. LDL Subfractions and Associations with Increased CVD Risk

Ip et al. published a systematic review of associations between LDL subfractions and cardiovascular outcomes in 2009 [38]. Most trials (37 of 52) found statistically significant associations between LDL subfraction size, number or phenotype, and cardiovascular outcomes. However, only 26 analyses were adjusted for standard lipids and just 12 of the adjusted studies found significant associations with incident cardiovascular disease, including 4 of the 14 trials analysing LDL size and sdLDL, and 8 of the 12 trials analysing LDL number and phenotype. Several important limitations were noted, including the lack of comparability between the various methods used to measure LDL subfractions and a heterogeneity in the adjustment and small sample sizes. The review concluded that there is a potential association between small LDL subfractions and incident cardiovascular disease, but there is not enough data to support their value as an independent risk factor for clinical use. Standardization of LDL subfractions measures is needed and studies need to confirm that treatment based on LDL subfraction testing is beneficial.

Since 2009, several large population-based longitudinal studies have investigated the association between various LDL subfractions and cardiovascular outcomes (Figure 2).

Mora et al. (2015) [24] published a study based on clinical data and blood samples from the JUPITER-study [39], a large, randomized prospective trial of rosuvastatin for primary CVD prevention in patients with LDL-C levels less than 130 mg/dL (<3.34 mmol/L) and elevated hsCRP. IM was used to determine the LDL subfractions. They included 11,186 participants, 5600 in the placebo arm, and 199 primary CVD events were recorded in the placebo group. Mean follow-up time was only 1.9 years as the trial was terminated early. In the placebo-arm of the study, several of the smaller LDL subfractions showed significant associations with an increased risk for incident CVD in a top vs.

bottom tertile analysis, which used a model that did not adjust for standard lipids (Figure 2). In a fully adjusted model including standard lipids, only one of the smaller LDL subfractions (LDL-Very small (c)) remained significantly associated with an increased CVD risk.

Shiffman et al. (2017) [25] and Melander et al. (2015) [40] published studies based on data and blood samples from the Malmö Prevention Project (MPP), a large population-based prospective study. Shiffman et al. included 5764 participants without CVD as a baseline, and recorded 1784 CVD events over a mean follow-up time of 8.05 years. Melander et al. included 1919 participants who were not candidates for statin therapy, as defined in the 2013 American Heart Association/American College of Cardiology guidelines for the treatment of cholesterol to reduce CVD risk, and, thus, were at low risk for a CVD event. During a mean follow-up of 16.2 years, they recorded 88 CVD events (4.8%). Both studies used IM as the method for the determination of LDL subfractions.

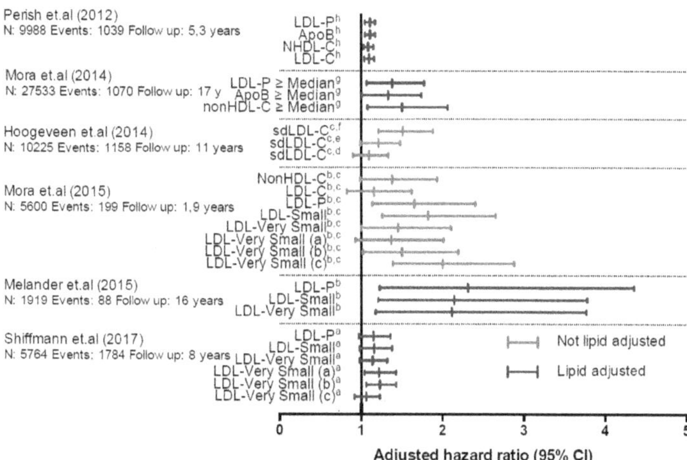

Figure 2. Associations between low-density lipoprotein (LDL) subfractions and cardiovascular disease (CVD) reported in prospective population-based studies published after 2009 and adjusted for confounding factors, such as age, smoking, gender, hypertension, etc. [a] Top vs. bottom quartile analysis. [b] Top vs. bottom tertile analysis. [c] Not significant in a model adjusted for lipids. [d] 1st vs. 2nd quartile analysis. [e] 1st vs. 3rd quartile analysis. [f] 1st vs. 4th quartile analysis. [g] Associations with CVD in patients with discordant levels of LDL particle number (LDL-P), apolipoprotein B (apoB), non-high density lipoprotein cholesterol (non-HDL-C) (≥median), and LDL cholesterol (LDL-C) (<median). [h] Hazard ratio (95% CI) pr. 1 SD higher. Hazard ratios for the outcome "major occlusive event" in the placebo arm are depicted in the graph.

Shiffman et al. reported that two of the five Small or Very Small LDL-subfractions were significantly associated with an increased risk for a CVD event in a top vs. bottom quartile analysis using a fully, lipid adjusted model. (Figure 2). LDL-P was not associated with an increased CVD risk. In a subgroup analysis of patients with low to intermediate risk, LDL-P and the LDL-Small subfraction was significantly associated with an increased risk, while in participants at very high risk (>20% 10-year risk), LDL-P and four of the five LDL subfractions were significantly associated with an increased risk.

Melander et al. reported that the LDL-P and the LDL-Small subfractions were associated with an increased risk of incident CVD in a fully, lipid adjusted, top vs. bottom tertile analysis, while the other four LDL-Very Small subfractions were not (Figure 2).

Hoogeveen et al. (2014) [41] published a study based on data and blood samples from the Atherosclerosis Risk in Communities (ARIC) study, a large population-based prospective study. They included 10,225 participants without prevalent CVD. Mean follow-up time was 11 years and incident CVD developed in 1158 participants. LDL subfractions were determined by a homogenous assay that quantified small, dense LDL-C (sdLDL-C). [16]. sdLDL-C was significantly associated with incident CVD in a top vs. bottom quartile analysis using a model that did not adjust for standard lipids (Figure 2), In a fully adjusted model that included standard lipids sdLDL-C was not associated with an increased risk. The cumulative incidence of CVD was higher in participants with low LDL-C and discordantly higher sdLDL-C (10.9%), compared to those with low sdLDL-C and discordantly higher LDL-C (7.9%). However, similar analyses using non-HDL-C, the recommended lipid metric for CVD risk assessment by most guidelines, was not reported.

Mora et al. (2014) [11] published a study based on the Women's Health Study, a large prospective population-based study. Participants were women aged >45 years and were free of cardiovascular disease at admission. This study included 27,533 participants, with a mean follow-up time of 17.2 years. During the follow-up period 1070 CVD-events occurred. The aim of this study was to determine the prevalence of and long-term prognosis for cardiovascular events in participants with discordant levels of LDL-C compared with LDL-P, non-HDL-C (NHDL-C), and apoB. LDL-P was determined using NMR. Among the women with below-median LDL-C and discordantly high NHDL-C, apoB, and LDL-P above the median, LDL-C underestimated the risk of cardiovascular events by 30–50% (Figure 2), but, importantly, LDL-P was not found to be superior to NHDH-C or apoB. This is in line with a 2009 study based on the same material [11].

Parish et al. (2012) [42] published a study based on the Heart Protection Study, a randomized controlled trial that included 20536 participants, between the ages of 40 and 80 years, who received either 40mg of simvastatin or a placebo once daily, with a mean follow up time of 5.3 years. Using baseline blood samples, they evaluated the associations between a variety of lipoprotein metrics, including LDL subfractions, determined by NMR, and cardiovascular outcomes. Primary endpoints (occlusive coronary events, revascularization, other cardiac events, and stroke) were equally strongly associated with LDL-C, apoB, NHDL-C, and LDL-P (Figure 2).

Overall, these studies seem to support an independent association between small LDL subfractions, or LDL-P, and hard cardiovascular outcomes, but inconsistencies between the studies should be noted and definitive evidence that LDL subfractions add predictive value to the established standard risk markers is lacking.

6. LDL Subfractions and PCSK9 Inhibitors

PCSK9 inhibitors reduce serum levels of LDL by blocking the degradation of the LDL receptor (LDLR) in hepatocytes and, thus, increasing the number of LDLRs on the cell surface, increasing LDL clearance in the liver [43]. Three major randomized trials have shown that PCSK9 inhibitors reduce the risk of cardiovascular events (FOURIER [44] and SPIRE-2 [45]) and all-cause mortality (ODYSSEY Outcomes—presented at ACC 2018, not published) in high-risk patients. The drugs are well tolerated and adverse effects are minimal, including no indications of adverse neurocognitive effects in patients with very low LDL-C levels. However, long-term data are needed [46].

Based on these results, PCSK9 inhibitors have now been included in CVD prevention guidelines [13,14] as a lipid lowering therapy to be considered for patients at very high risk of CVD. Very high risk patients are usually defined as patients with familial hypercholesterolemia (FH), in secondary prevention for patients with high residual risk after statin therapy, or patients who do not tolerate appropriate doses of statins. Currently, individuals without FH in primary prevention do not fulfil cost-effectiveness criteria due to the high costs of these new drugs [47,48].

Pre-clinical trials have reported an independent association between plasma levels of PCSK9 and small LDL subfractions in patients with established CVD [49,50] and abdominally obese dyslipidemic men [51], but not in healthy subjects [52], suggesting that plasma levels of PCSK9 could be related to the metabolism of small LDL subfractions in patients at risk of CVD. An impaired affinity of sdLDL to LDLR [33] could suggest that PCSK9 inhibitors are less efficient in reducing the small LDL particles compared to the larger LDL particles. As the PCSK9 inhibitors have just recently been introduced, a limited number of clinical studies evaluating the effects of PCSK9 inhibitors on LDL subfractions have been published. Baruch et al. published a study on the effects of RG7652, a novel PCSK9 inhibitor, on LDL subfractions evaluated by NMR in a randomized trial including 248 patients [53]. In this trial, they demonstrated reductions in LDL-C of up to 52.4%, which is comparable to other PCSK9 inhibitors that are already available. Both the small and large LDL-particles were reduced significantly at the optimal dose of RG7652, but the median percentage change was lower for the smaller LDL-particles compared to the larger particles (−43% vs. −81% from baseline), and 11 of the 45 patients showed an increase in the small LDL particles at the end of the study. Two smaller studies on the PCSK9 inhibitors, alirocumab and evolocumab, also reported a proportionally larger reduction in the large LDL particles compared to the small LDL particles, but significant reductions nonetheless [54,55]. In a pilot study of three patients with FH, who converted from lipid apheresis to evolocumab, our group reported reductions of all LDL subfractions [56]. For comparison, studies evaluating the effects of statins on LDL subfractions are more discordant, with no changes [57,58], significant decreases [59–62], and also slight increases [63] of LDL subfraction metrics being reported.

To summarize, PCSK9 inhibitors seem to be effective at lowering all LDL subfractions, but with a trend towards a more efficient lowering of the larger LDL subfractions.

7. Conclusions

PCKS9 inhibitors have been proven to reduce the risk of cardiovascular events and all-cause mortality in patients at high risk of CVD, with minimal adverse effects reported in short-term trials. With these new drugs, physicians now have another powerful tool to aggressively target and lower LDL-C, a causal agent in the development of atherosclerotic cardiovascular disease. As these drugs come at a considerable cost, and with long-term adverse effects unknown, it is important to critically evaluate when patients should be considered for PCKS9 inhibitors. Current CVD prevention guidelines have stated that patients at very high risk of CVD should be considered for PCSK9 inhibitors.

It has been proposed that measuring LDL subfractions, specifically small dense LDL or LDL-P, could improve CVD risk assessment and identify patients at high risk of CVD. In-vitro studies have indicated that small, dense LDLs are more atherogenic than larger LDL particles. Several methods for the determination of LDL subfractions and LDL-P are used in the published literature, however, several factors limit their availability. Furthermore, a lack of standardization makes comparison between these methods challenging. As these methods separate LDL particles based on completely different physicochemical properties, it is not possible to determine whether the different methods are measuring the same subfractions.

Longitudinal population studies indicate that increased levels of small dense LDL, or increased LDL-P, are independently associated with an increased risk of cardiovascular disease. However, definitive evidence that LDL subfractions, or LDL-P, add predictive value to the standard risk markers, which are readily available at most clinical laboratories, is lacking. No studies have shown that these measurements improve clinical outcomes. Based on the 2009 American Heart Association (AHA) framework for novel risk markers [64], a new risk marker should predict future outcomes in prospective studies, add predictive information to established risk markers, improve clinical outcomes, and be determined as cost-effective when compared to established risk markers. This framework sets a high standard for novel risk markers and should be considered a guideline. However, evidence that the novel risk marker adds predictive value to an established marker is of critical importance.

Despite indications that PCSK9 inhibitors are effective at lowering sdLDL and LDL-P, important pieces of evidence proving that LDL subfractions and LDL-P yield clinically useful information is lacking, and these measurements are not recommended in the evaluation of whether to initiate PCKS9 inhibitors in patients at risk of CVD.

In order to use analysis of LDL subfractions in clinical risk evaluation, standardization of methods and a general agreement regarding which fractions to include in such an assessment is required.

Author Contributions: C.A.K., A.H., and K.T.L. contributed to the review concept and design. C.A.K., A.H., and K.T.L. drafted the manuscript. All authors made critical revision of the article for important intellectual content.

Conflicts of Interest: The authors declare no conflicts of interest.

References

1. WHO. *Cardiovascular Disease*; WHO: Geneva, Switzerland; Available online: http://www.who.int/cardiovascular_diseases/en/ (accessed on 26 May 2018).
2. Morris, P.B.; Ballantyne, C.M.; Birtcher, K.K.; Dunn, S.P.; Urbina, E.M. Review of clinical practice guidelines for the management of LDL-related risk. *J. Am. Coll. Cardiol.* **2014**, *64*, 196–206. [CrossRef] [PubMed]
3. Tibrewala, A.; Jivan, A.; Oetgen, W.J.; Stone, N.J. A comparative analysis of current lipid treatment guidelines: Nothing stands still. *J. Am. Coll. Cardiol.* **2018**, *71*, 794–799. [CrossRef] [PubMed]
4. Yusuf, S.; Hawken, S.; Ounpuu, S.; Dans, T.; Avezum, A.; Lanas, F.; McQueen, M.; Budaj, A.; Pais, P.; Varigos, J.; et al. Effect of potentially modifiable risk factors associated with myocardial infarction in 52 countries (the interheart study): Case-control study. *Lancet* **2004**, *364*, 937–952. [CrossRef]
5. Ference, B.A.; Ginsberg, H.N.; Graham, I.; Ray, K.K.; Packard, C.J.; Bruckert, E.; Hegele, R.A.; Krauss, R.M.; Raal, F.J.; Schunkert, H.; et al. Low-density lipoproteins cause atherosclerotic cardiovascular disease 1. Evidence from genetic, epidemiologic, and clinical studies. A consensus statement from the european atherosclerosis society consensus panel. *Eur. Heart J.* **2017**, *38*, 2459–2472. [CrossRef] [PubMed]
6. Piepoli, M.F.; Hoes, A.W.; Agewall, S.; Albus, C.; Brotons, C.; Catapano, A.L.; Cooney, M.T.; Corra, U.; Cosyns, B.; Deaton, C.; et al. 2016 european guidelines on cardiovascular disease prevention in clinical practice. The sixth joint task force of the european society of cardiology and other societies on cardiovascular disease prevention in clinical practice.(constituted by representatives of 10 societies and by invited experts. Developed with the special contribution of the european association for cardiovascular prevention & rehabilitation. *Giorn. Ital. Cardiol. (Rome)* **2017**, *18*, 547–612. [PubMed]
7. Ioannidis, J.P.A. Inconsistent guideline recommendations for cardiovascular prevention and the debate about zeroing in on and zeroing LDL-C levels with PCSK9 inhibitors. *JAMA* **2017**, *318*, 419–420. [CrossRef] [PubMed]
8. Martin, S.S.; Blaha, M.J.; Elshazly, M.B.; Brinton, E.A.; Toth, P.P.; McEvoy, J.W.; Joshi, P.H.; Kulkarni, K.R.; Mize, P.D.; Kwiterovich, P.O.; et al. Friedewald-estimated versus directly measured low-density lipoprotein cholesterol and treatment implications. *J. Am. Coll. Cardiol.* **2013**, *62*, 732–739. [CrossRef] [PubMed]
9. Sachdeva, A.; Cannon, C.P.; Deedwania, P.C.; Labresh, K.A.; Smith, S.C., Jr.; Dai, D.; Hernandez, A.; Fonarow, G.C. Lipid levels in patients hospitalized with coronary artery disease: An analysis of 136,905 hospitalizations in get with the guidelines. *Am. Heart J.* **2009**, *157*, 111–117. [CrossRef] [PubMed]
10. Otvos, J.D.; Mora, S.; Shalaurova, I.; Greenland, P.; Mackey, R.H.; Goff, D.C., Jr. Clinical implications of discordance between low-density lipoprotein cholesterol and particle number. *J. Clin. Lipidol.* **2011**, *5*, 105–113. [CrossRef] [PubMed]
11. Mora, S.; Buring, J.E.; Ridker, P.M. Discordance of low-density lipoprotein (LDL) cholesterol with alternative ldl-related measures and future coronary events. *Circulation* **2014**, *129*, 553–561. [CrossRef] [PubMed]
12. McGettigan, P.; Ferner, R.E. PCSK9 inhibitors for hypercholesterolaemia. *BMJ* **2017**, *356*, j188. [CrossRef] [PubMed]
13. Landmesser, U.; Chapman, M.J.; Stock, J.K.; Amarenco, P.; Belch, J.J.F.; Boren, J.; Farnier, M.; Ference, B.A.; Gielen, S.; Graham, I.; et al. 2017 update of ESC/EAS task force on practical clinical guidance for proprotein convertase subtilisin/kexin type 9 inhibition in patients with atherosclerotic cardiovascular disease or in familial hypercholesterolaemia. *Eur. Heart J.* **2018**, *39*, 1131–1140. [CrossRef] [PubMed]

14. Lloyd-Jones, D.M.; Morris, P.B.; Ballantyne, C.M.; Birtcher, K.K.; Daly, D.D., Jr.; DePalma, S.M.; Minissian, M.B.; Orringer, C.E.; Smith, S.C., Jr. 2017 focused update of the 2016 ACC expert consensus decision pathway on the role of non-statin therapies for LDL-cholesterol lowering in the management of atherosclerotic cardiovascular disease risk: A report of the american college of cardiology task force on expert consensus decision pathways. *J. Am. Coll. Cardiol.* **2017**, *70*, 1785–1822. [PubMed]
15. Ramasamy, I. Update on the laboratory investigation of dyslipidemias. *Clin. Chim. Acta* **2018**, *479*, 103–125. [CrossRef] [PubMed]
16. Ito, Y.; Fujimura, M.; Ohta, M.; Hirano, T. Development of a homogeneous assay for measurement of small dense LDL cholesterol. *Clin. Chem.* **2011**, *57*, 57–65. [CrossRef] [PubMed]
17. Hoefner, D.M.; Hodel, S.D.; O'Brien, J.F.; Branum, E.L.; Sun, D.; Meissner, I.; McConnell, J.P. Development of a rapid, quantitative method for LDL subfractionation with use of the quantimetrix lipoprint LDL system. *Clin. Chem.* **2001**, *47*, 266–274. [PubMed]
18. Lipoprint, Q.L. Available online: https://quantimetrix.com/lipoprint/lipoprint-profile/ (accessed on 26 May 2018).
19. LipoProfile, N. Available online: http://www.clevelandheartlab.com/tests/nmr-lipoprofile/ (accessed on 26 May 2018).
20. Fractionation, Q.D.C.I. Available online: https://education.questdiagnostics.com/faq/FAQ134 (accessed on 26 May 2018).
21. Sniderman, A.; Shapiro, S.; Marpole, D.; Skinner, B.; Teng, B.; Kwiterovich, P.O., Jr. Association of coronary atherosclerosis with hyperapobetalipoproteinemia [increased protein but normal cholesterol levels in human plasma low density (beta) lipoproteins]. *Proc. Natl. Acad. Sci. USA* **1980**, *77*, 604–608. [CrossRef] [PubMed]
22. Austin, M.A.; King, M.C.; Vranizan, K.M.; Newman, B.; Krauss, R.M. Inheritance of low-density lipoprotein subclass patterns: Results of complex segregation analysis. *Am. J. Hum. Genet.* **1988**, *43*, 838–846. [PubMed]
23. Chung, M.; Lichtenstein, A.H.; Ip, S.; Lau, J.; Balk, E.M. Comparability of methods for LDL subfraction determination: A systematic review. *Atherosclerosis* **2009**, *205*, 342–348. [CrossRef] [PubMed]
24. Mora, S.; Caulfield, M.P.; Wohlgemuth, J.; Chen, Z.; Superko, H.R.; Rowland, C.M.; Glynn, R.J.; Ridker, P.M.; Krauss, R.M. Atherogenic lipoprotein subfractions determined by ion mobility and first cardiovascular events after random allocation to high-intensity statin or placebo: The justification for the use of statins in prevention: An intervention trial evaluating rosuvastatin (jupiter) trial. *Circulation* **2015**, *132*, 2220–2229. [PubMed]
25. Shiffman, D.; Louie, J.Z.; Caulfield, M.P.; Nilsson, P.M.; Devlin, J.J.; Melander, O. LDL subfractions are associated with incident cardiovascular disease in the malmo prevention project study. *Atherosclerosis* **2017**, *263*, 287–292. [CrossRef] [PubMed]
26. Hopkins, P.N.; Pottala, J.V.; Nanjee, M.N. A comparative study of four independent methods to measure LDL particle concentration. *Atherosclerosis* **2015**, *243*, 99–106. [CrossRef] [PubMed]
27. Sninsky, J.J.; Rowland, C.M.; Baca, A.M.; Caulfield, M.P.; Superko, H.R. Classification of LDL phenotypes by 4 methods of determining lipoprotein particle size. *J. Investig. Med.* **2013**, *61*, 942–949. [CrossRef] [PubMed]
28. Williams, P.T.; Zhao, X.Q.; Marcovina, S.M.; Otvos, J.D.; Brown, B.G.; Krauss, R.M. Comparison of four methods of analysis of lipoprotein particle subfractions for their association with angiographic progression of coronary artery disease. *Atherosclerosis* **2014**, *233*, 713–720. [CrossRef] [PubMed]
29. Robinson, J.G. What is the role of advanced lipoprotein analysis in practice? *J. Am. Coll. Cardiol.* **2012**, *60*, 2607–2615. [CrossRef] [PubMed]
30. Tabas, I.; Williams, K.J.; Boren, J. Subendothelial lipoprotein retention as the initiating process in atherosclerosis: Update and therapeutic implications. *Circulation* **2007**, *116*, 1832–1844. [CrossRef] [PubMed]
31. Goldstein, J.L.; Brown, M.S. A century of cholesterol and coronaries: From plaques to genes to statins. *Cell* **2015**, *161*, 161–172. [CrossRef] [PubMed]
32. Ivanova, E.A.; Myasoedova, V.A.; Melnichenko, A.A.; Grechko, A.V.; Orekhov, A.N. Small dense low-density lipoprotein as biomarker for atherosclerotic diseases. *Oxid. Med. Cell. Longev.* **2017**, *2017*, 1273042. [CrossRef] [PubMed]
33. Nigon, F.; Lesnik, P.; Rouis, M.; Chapman, M.J. Discrete subspecies of human low density lipoproteins are heterogeneous in their interaction with the cellular LDL receptor. *J. Lipid Res.* **1991**, *32*, 1741–1753. [PubMed]

34. Tribble, D.L.; Rizzo, M.; Chait, A.; Lewis, D.M.; Blanche, P.J.; Krauss, R.M. Enhanced oxidative susceptibility and reduced antioxidant content of metabolic precursors of small, dense low-density lipoproteins. *Am. J. Med.* **2001**, *110*, 103–110. [CrossRef]

35. Tertov, V.V.; Kaplun, V.V.; Sobenin, I.A.; Boytsova, E.Y.; Bovin, N.V.; Orekhov, A.N. Human plasma trans-sialidase causes atherogenic modification of low density lipoprotein. *Atherosclerosis* **2001**, *159*, 103–115. [CrossRef]

36. Berneis, K.K.; Krauss, R.M. Metabolic origins and clinical significance of LDL heterogeneity. *J. Lipid Res.* **2002**, *43*, 1363–1379. [CrossRef] [PubMed]

37. Chapman, M.J.; Guerin, M.; Bruckert, E. Atherogenic, dense low-density lipoproteins. Pathophysiology and new therapeutic approaches. *Eur. Heart J.* **1998**, *19*, A24–A30. [PubMed]

38. Ip, S.; Lichtenstein, A.H.; Chung, M.; Lau, J.; Balk, E.M. Systematic review: Association of low-density lipoprotein subfractions with cardiovascular outcomes. *Ann. Intern Med.* **2009**, *150*, 474–484. [CrossRef] [PubMed]

39. Ridker, P.M.; Danielson, E.; Fonseca, F.A.; Genest, J.; Gotto, A.M., Jr.; Kastelein, J.J.; Koenig, W.; Libby, P.; Lorenzatti, A.J.; MacFadyen, J.G.; et al. Rosuvastatin to prevent vascular events in men and women with elevated c-reactive protein. *N. Engl. J. Med.* **2008**, *359*, 2195–2207. [CrossRef] [PubMed]

40. Melander, O.; Shiffman, D.; Caulfield, M.P.; Louie, J.Z.; Rowland, C.M.; Devlin, J.J.; Krauss, R.M. Low-density lipoprotein particle number is associated with cardiovascular events among those not classified into statin benefit groups. *J. Am. Coll. Cardiol.* **2015**, *65*, 2571–2573. [CrossRef] [PubMed]

41. Hoogeveen, R.C.; Gaubatz, J.W.; Sun, W.; Dodge, R.C.; Crosby, J.R.; Jiang, J.; Couper, D.; Virani, S.S.; Kathiresan, S.; Boerwinkle, E.; et al. Small dense low-density lipoprotein-cholesterol concentrations predict risk for coronary heart disease: The atherosclerosis risk in communities (ARIC) study. *Arterioscler. Thromb. Vasc. Biol.* **2014**, *34*, 1069–1077. [CrossRef] [PubMed]

42. Parish, S.; Offer, A.; Clarke, R.; Hopewell, J.C.; Hill, M.R.; Otvos, J.D.; Armitage, J.; Collins, R.; Heart Protection Study Collaborative Group. Lipids and lipoproteins and risk of different vascular events in the mrc/bhf heart protection study. *Circulation* **2012**, *125*, 2469–2478. [CrossRef] [PubMed]

43. Lagace, T.A. PCSK9 and ldlr degradation: Regulatory mechanisms in circulation and in cells. *Curr. Opin. Lipidol.* **2014**, *25*, 387–393. [CrossRef] [PubMed]

44. Sabatine, M.S.; Giugliano, R.P.; Keech, A.C.; Honarpour, N.; Wiviott, S.D.; Murphy, S.A.; Kuder, J.F.; Wang, H.; Liu, T.; Wasserman, S.M.; et al. Evolocumab and clinical outcomes in patients with cardiovascular disease. *N. Engl. J. Med.* **2017**, *376*, 1713–1722. [CrossRef] [PubMed]

45. Ridker, P.M.; Revkin, J.; Amarenco, P.; Brunell, R.; Curto, M.; Civeira, F.; Flather, M.; Glynn, R.J.; Gregoire, J.; Jukema, J.W.; et al. Cardiovascular efficacy and safety of bococizumab in high-risk patients. *N. Engl. J. Med.* **2017**, *376*, 1527–1539. [CrossRef] [PubMed]

46. Giugliano, R.P.; Mach, F.; Zavitz, K.; Kurtz, C.; Im, K.; Kanevsky, E.; Schneider, J.; Wang, H.; Keech, A.; Pedersen, T.R.; et al. Cognitive function in a randomized trial of evolocumab. *N. Engl. J. Med.* **2017**, *377*, 633–643. [CrossRef] [PubMed]

47. Korman, M.; Wisloff, T. Modelling the cost-effectiveness of pcsk9 inhibitors vs. Ezetimibe through ldl-c reductions in a norwegian setting. *Eur. Heart J. Cardiovasc. Pharmacother.* **2018**, *4*, 15–22. [CrossRef] [PubMed]

48. Bonow, R.O.; Harrington, R.A.; Yancy, C.W. Cost-effectiveness of pcsk9 inhibitors: Proof in the modeling. *JAMA Cardiol.* **2017**, *2*, 1298–1299. [CrossRef] [PubMed]

49. Xu, R.X.; Li, S.; Zhang, Y.; Li, X.L.; Guo, Y.L.; Zhu, C.G.; Li, J.J. Relation of plasma pcsk9 levels to lipoprotein subfractions in patients with stable coronary artery disease. *Lipids Health Dis.* **2014**, *13*, 188. [CrossRef] [PubMed]

50. Zhang, Y.; Xu, R.X.; Li, S.; Zhu, C.G.; Guo, Y.L.; Sun, J.; Li, J.J. Association of plasma small dense LDL cholesterol with PCSK9 levels in patients with angiographically proven coronary artery disease. *Nutr. Metab. Cardiovasc. Dis.* **2015**, *25*, 426–433. [CrossRef] [PubMed]

51. Arsenault, B.J.; Pelletier-Beaumont, E.; Almeras, N.; Tremblay, A.; Poirier, P.; Bergeron, J.; Despres, J.P. PCSK9 levels in abdominally obese men: Association with cardiometabolic risk profile and effects of a one-year lifestyle modification program. *Atherosclerosis* **2014**, *236*, 321–326. [CrossRef] [PubMed]

52. Kwakernaak, A.J.; Lambert, G.; Dullaart, R.P. Plasma proprotein convertase subtilisin-kexin type 9 is predominantly related to intermediate density lipoproteins. *Clin. Biochem.* **2014**, *47*, 679–682. [CrossRef] [PubMed]

53. Baruch, A.; Mosesova, S.; Davis, J.D.; Budha, N.; Vilimovskij, A.; Kahn, R.; Peng, K.; Cowan, K.J.; Harris, L.P.; Gelzleichter, T.; et al. Effects of rg7652, a monoclonal antibody against PCSK9, on LDL-c, LDL-c subfractions, and inflammatory biomarkers in patients at high risk of or with established coronary heart disease (from the phase 2 equator study). *Am. J. Cardiol.* **2017**, *119*, 1576–1583. [CrossRef] [PubMed]

54. Koren, M.J.; Kereiakes, D.; Pourfarzib, R.; Winegar, D.; Banerjee, P.; Hamon, S.; Hanotin, C.; McKenney, J.M. Effect of PCSK9 inhibition by alirocumab on lipoprotein particle concentrations determined by nuclear magnetic resonance spectroscopy. *J. Am. Heart Assoc.* **2015**, *4*, e002224. [PubMed]

55. Wu, L.; Bamberger, C.; Waldmann, E.; Stark, R.; Henze, K.; Parhofer, K. The effect of PCSK9 inhibition on LDL-subfractions in patients with severe ldl-hypercholesterolemia. *J. Am. Coll. Cardiol.* **2017**, *69*, 1719. [CrossRef]

56. Lappegard, K.T.; Kjellmo, C.A.; Ljunggren, S.; Cederbrant, K.; Marcusson-Stahl, M.; Mathisen, M.; Karlsson, H.; Hovland, A. Lipoprotein apheresis affects lipoprotein particle subclasses more efficiently compared to the PCSK9 inhibitor evolocumab, a pilot study. *Transfus. Apher. Sci.* **2018**, *57*, 91–96. [CrossRef] [PubMed]

57. Blake, G.J.; Albert, M.A.; Rifai, N.; Ridker, P.M. Effect of pravastatin on LDL particle concentration as determined by NMR spectroscopy: A substudy of a randomized placebo controlled trial. *Eur. Heart J.* **2003**, *24*, 1843–1847. [CrossRef] [PubMed]

58. Superko, H.R.; Krauss, R.M.; DiRicco, C. Effect of fluvastatin on low-density lipoprotein peak particle diameter. *Am. J. Cardiol.* **1997**, *80*, 78–81. [CrossRef]

59. Frost, R.J.; Otto, C.; Geiss, H.C.; Schwandt, P.; Parhofer, K.G. Effects of atorvastatin versus fenofibrate on lipoprotein profiles, low-density lipoprotein subfraction distribution, and hemorheologic parameters in type 2 diabetes mellitus with mixed hyperlipoproteinemia. *Am. J. Cardiol.* **2001**, *87*, 44–48. [CrossRef]

60. Otvos, J.D.; Shalaurova, I.; Freedman, D.S.; Rosenson, R.S. Effects of pravastatin treatment on lipoprotein subclass profiles and particle size in the PLAC-I trial. *Atherosclerosis* **2002**, *160*, 41–48. [CrossRef]

61. Karalis, D.G.; Ishisaka, D.Y.; Luo, D.; Ntanios, F.; Wun, C.C. Effects of increasing doses of atorvastatin on the atherogenic lipid subclasses commonly associated with hypertriglyceridemia. *Am. J. Cardiol.* **2007**, *100*, 445–449. [CrossRef] [PubMed]

62. Chapman, M.J.; Orsoni, A.; Robillard, P.; Therond, P.; Giral, P. Duality of statin action on lipoprotein subpopulations in the mixed dyslipidemia of metabolic syndrome: Quantity vs. quality over time and implication of cetp. *J. Clin. Lipidol.* **2018**. [CrossRef] [PubMed]

63. Soedamah-Muthu, S.S.; Colhoun, H.M.; Thomason, M.J.; Betteridge, D.J.; Durrington, P.N.; Hitman, G.A.; Fuller, J.H.; Julier, K.; Mackness, M.I.; Neil, H.A.; et al. The effect of atorvastatin on serum lipids, lipoproteins and NMR spectroscopy defined lipoprotein subclasses in type 2 diabetic patients with ischaemic heart disease. *Atherosclerosis* **2003**, *167*, 243–255. [CrossRef]

64. Hlatky, M.A.; Greenland, P.; Arnett, D.K.; Ballantyne, C.M.; Criqui, M.H.; Elkind, M.S.; Go, A.S.; Harrell, F.E., Jr.; Hong, Y.; Howard, B.V.; et al. Criteria for evaluation of novel markers of cardiovascular risk: A scientific statement from the american heart association. *Circulation* **2009**, *119*, 2408–2416. [CrossRef] [PubMed]

 *diseases*

Review

Management of Dyslipidemia in Type 2 Diabetes: Recent Advances in Nonstatin Treatment

Kazutoshi Sugiyama and Yoshifumi Saisho *

Division of Endocrinology, Metabolism, and Nephrology, Department of Internal Medicine,
Keio University School of Medicine, 35 Shinanomachi, Shinjuku-ku, Tokyo 160-8582, Japan; ksugiyama@keio.jp
* Correspondence: ysaisho@keio.jp; Tel.: +81-3-5363-3797

Received: 3 May 2018; Accepted: 22 May 2018; Published: 24 May 2018

Abstract: Dyslipidemia is a major risk factor for cardiovascular disease (CVD), which is the leading cause of morbidity and mortality in type 2 diabetes (T2DM). Statins have played a crucial role in its management, but residual risk remains since many patients cannot achieve their desired low-density lipoprotein cholesterol (LDL-C) level and up to 20% of patients are statin-intolerant, experiencing adverse events perceived to be caused by statins, most commonly muscle symptoms. Recently, great advances have been made in nonstatin treatment with ezetimibe, a cholesterol absorption inhibitor, and proprotein convertase subtilisin/kexin type 9 (PCSK9) monoclonal antibodies (mAbs), all showing a proven benefit with an excellent safety profile in cardiovascular outcome trials. This review summarizes the key aspects and the evolving role of these agents in the management of dyslipidemia in patients with T2DM, along with a brief introduction of novel drugs currently in development.

Keywords: dyslipidemia; type 2 diabetes; PCSK9; nonstatin

---

1. Introduction

Cardiovascular disease (CVD) is the leading cause of morbidity and mortality in individuals with type 2 diabetes (T2DM) [1,2]. In the Finnish study, which showed that T2DM is a coronary heart disease (CHD) equivalent, type 2 diabetic patients without prior myocardial infarction (MI) and nondiabetic patients with prior MI had a similar incidence of MI and risk of CHD death. Furthermore, when less stringent criteria for prior CHD (MI, angina pectoris, or ischemic electrocardiogram changes) were used, T2DM carried a larger risk than prior CHD [3,4]. Diabetic dyslipidemia, characterized by increased triglyceride (TG) level and decreased high-density lipoprotein cholesterol (HDL-C) level, is a major risk factor for CVD. Although low-density lipoprotein cholesterol (LDL-C) level is typically normal, small dense LDL particles, which are more atherogenic since they are more likely to undergo glycation and oxidation, are more prevalent in T2DM [5].

CARDS (Collaborative Atorvastatin Diabetes Study) [6] and the Heart Protection Study [7] have shown the efficacy of statins in diabetic patients, and in the Cholesterol Treatment Trialists' meta-analysis of diabetic patients, statins reduced major vascular events by 21% and all-cause mortality by 9% for each 38.7 mg/dL (1.0 mmol/L) reduction in LDL-C [8]. Major guidelines recommend statins for the treatment of dyslipidemia in T2DM [2,9–12], but even with high-dose statins, 12.7% and 40.4% of patients do not achieve LDL-C levels below 100 mg/dL and 70 mg/dL, respectively [13]. Moreover, in clinical practice, statin-associated muscle symptoms occur in up to 20% of patients and contribute to their discontinuation [14]. New treatment strategies are needed, and this review focuses on recent advances in nonstatin treatment, with special attention to proprotein convertase subtilisin/kexin type 9 (PCSK9) monoclonal antibodies (mAbs).

2. Search Strategy

We searched PubMed, ClinicalTrials.gov, and other sources for articles published in English between 1 January 2000 and 1 April 2018, using the search terms "dyslipidemia", "type 2 diabetes", "PCSK9", "evolocumab", and "alirocumab". We also searched the reference lists of articles identified by this search strategy along with manually selected articles known to the authors.

3. Nonstatin Lipid-Lowering Therapies

3.1. Ezetimibe

Ezetimibe reduces cholesterol absorption by inhibiting Niemann-Pick C1-Like 1 (NPC1L1) protein in the small intestine and hepatocytes [15,16]. In the IMPROVE-IT (Improved Reduction of outcomes: Vytorin Efficacy International Trial) [17], the first trial to show an improvement in cardiovascular (CV) outcomes with the addition of a nonstatin drug to a statin, 18,144 patients who had been hospitalized for an acute coronary syndrome (ACS) within the preceding 10 days were randomized to simvastatin–ezetimibe combination therapy or simvastatin monotherapy. With a median follow-up of 6 years, the addition of ezetimibe to simvastatin reduced LDL-C by 16 mg/dL and resulted in a 6.4% reduction (32.7% vs. 34.7%; hazard ratio (HR) 0.936; 95% confidence interval (CI) 0.89–0.99; $p = 0.016$) in the primary endpoint, which was a composite of CV death, MI, unstable angina requiring hospitalization, coronary revascularization, or stroke, compared to simvastatin monotherapy (Table 1). There were no differences in adverse events, including muscle-related events.

Table 1. Cardiovascular outcome trials of nonstatin drugs.

Variable	IMPROVE-IT [17]	FOURIER [18]	ODYSSEY Outcomes [19]
No. of patients	18,144	27,564	18,924
No. of patients with diabetes	4933 (27%)	11,031 (40%) [20]	5444 (29%)
Mean age (years)	64	63	58
Clinical characteristics	ACS within 10 days	ASCVD and LDL-C $\geq$70 mg/dL or non-HDL-C $\geq$100 mg/dL on statin	ACS within 12 months; LDL-C $\geq$70 mg/dL or non-HDL-C $\geq$100 mg/dL or ApoB $\geq$80 mg/dL on high-intensity statin
Intervention	Simvastatin 40 mg and ezetimibe 10 mg vs. simvastatin 40 mg	Evolocumab 140 mg q 2w or 420 mg q 4w vs. placebo	Alirocumab 75–150 mg q 2w vs. placebo
Primary endpoint	CV death, MI, stroke, hospitalization for UA, coronary revascularization	CV death, MI, stroke, hospitalization for UA, coronary revascularization	CHD death, MI, ischemic stroke, hospitalization for UA
Median f/u (years)	6	2.2	2.8
Achieved LDL-C (mg/dL)	53.7 vs. 69.5	30 vs. 92	53.3 vs. 101.4
Primary endpoint	32.7% vs. 34.7%; HR 0.936 (95% CI 0.89–0.99); $p = 0.016$	9.8% vs. 11.3%; HR 0.85 (95% CI 0.79–0.92); $p < 0.001$	9.5% vs. 11.1%; HR 0.85 (95% CI 0.78–0.93); $p = 0.0003$
3-point MACE (CV death, MI, stroke)	22.2% vs. 20.4%; HR 0.90 (95% CI 0.84–0.96); $p = 0.003$	5.9% vs. 7.4%; HR 0.80 (95% CI 0.73–0.88); $p<0.001$	10.3% vs. 11.9%; HR 0.86 (95% CI 0.79–0.93); $p = 0.0003$ *
CV death	6.8% vs. 6.9%; HR 1.00 (95% CI 0.89–1.13); $p = 1.00$	1.8% vs. 1.7%; HR 1.05 (95% CI 0.88–1.25); $p = 0.62$	2.5% vs. 2.9%; HR 0.88 (95% CI 0.74–1.05); $p = 0.15$
All-cause death	15.3% vs. 15.4%; HR 0.99 (95% CI 0.91–1.07); $p = 0.78$	3.2% vs. 3.1%; HR 1.04 (95% CI 0.91–1.19); $p = 0.54$	3.5% vs. 4.1%; HR 0.85 (95% CI 0.73–0.98); $p = 0.026$
Adverse events	Similar safety in both groups	Injection-site reactions: 2.1% vs. 1.6% Neutralizing antibodies: 0% in both groups	Injection site reactions: 3.8% vs. 2.1% Neutralizing antibodies: 0.4% vs. 0.1%

ACS = acute coronary syndrome; AMI = acute myocardial infarction; ApoB = apolipoprotein B; ASCVD = atherosclerotic cardiovascular disease; CHD = coronary heart disease; CI = confidence interval; CV = cardiovascular; FOURIER = Further Cardiovascular Outcomes Research with PCSK9 Inhibition in Subjects with Elevated Risk; HR = hazard ratio; HDL-C = high-density lipoprotein cholesterol; IMPROVE-IT = Improved Reduction of outcomes: Vytorin Efficacy International Trial; LDL-C = low-density lipoprotein cholesterol; MACE = major adverse cardiovascular events; MI = myocardial infarction; UA = unstable angina; * 3-point MACE for all-cause death, MI, stroke.

Of the study subjects, 4933 (27%) had diabetes at baseline, and compared to patients without diabetes, ezetimibe was associated with an enhanced benefit ((HR 0.86; 95% CI 0.78–0.94 for patients with diabetes) vs. (HR 0.98; 95% CI 0.92–1.04 for patients without diabetes); p value = 0.023 for interaction) in the primary endpoint with similar safety outcomes [21]. In addition, in a prespecified analysis which compared outcomes stratified by achieved LDL-C level at 1 month, the adjusted HRs for the primary endpoint favored lower achieved LDL-C groups (HRs of 1.0, 0.82, 0.80, and 0.79 for LDL-C >70, 50–69, 30–49, and <30 mg/dL, respectively; p for trend <0.001) without an increase in adverse events [22]. These results suggest the benefit of the addition of ezetimibe in diabetic patients after an ACS and perhaps also in patients with stable clinical atherosclerotic cardiovascular disease (ASCVD).

3.2. PCSK9 Inhibitors

In 2003, gain-of-function mutations in PCSK9 were reported as a cause of hypercholesterolemia [23]. Soon after, low LDL-C in individuals with loss-of-function mutations in PCSK9 was reported [24], and a moderate lifelong reduction in LDL-C (15–28%) resulted in a substantial reduction in the incidence of CHD by 47–88% [25]. Furthermore, individuals with no circulating PCSK9 and very low LDL-C due to compound heterozygous loss-of-function mutations in PCSK9 were apparently healthy [26], making PCSK9 inhibition a very attractive target for LDL-C-lowering therapy.

Plasma LDL-C binds to LDL receptors expressed on the surface of hepatocytes and is internalized by endocytosis [27]. LDL receptors are usually recycled to the cell surface, but when PCSK9 binds with LDL receptors, LDL receptors are delivered to lysosomes for degradation, resulting in lower expression of LDL receptors and an increase in LDL-C [28]. Therapeutic approaches targeting extracellular PCSK9 (e.g., mAbs) and intracellular PCSK9 (e.g., small interfering RNAs (siRNAs)) are currently under investigation, but mAbs have been the most successful strategy thus far [29].

3.2.1. Monoclonal Antibodies

Evolocumab and alirocumab, two subcutaneous agents currently available on the market, have been studied in numerous populations, including familial hypercholesterolemia, diabetes, and statin intolerance, as monotherapy and in combination with statins. In a meta-analysis of phase 2 and 3 studies, treatment with PCSK9 mAbs reduced LDL-C by 55% [30]. Moreover, OSLER (Open-Label Study of Long-Term Evaluation against LDL Cholesterol) 1 and 2 [31] and ODYSSEY LONG TERM (Long-term Safety and Tolerability of Alirocumab in High Cardiovascular Risk Patients with Hypercholesterolemia Not Adequately Controlled with Their Lipid-Modifying Therapy) [32], which were long-term studies of 1 to 1.5 years, have shown a significant reduction in CV events of roughly 50%. However, the number of CV events was small, and confirmation in trials adequately powered to examine CV outcomes are eagerly awaited. The results of FOURIER (Further Cardiovascular Outcomes Research with PCSK9 Inhibition in Subjects with Elevated Risk) [18] and ODYSSEY Outcomes [19] have been recently reported and are discussed below.

3.2.2. Cardiovascular Outcomes Trial: FOURIER

FOURIER [18] included 27,564 patients with ASCVD and LDL-C $\geq$ 70 mg/dL who were receiving statin therapy. Patients were randomly assigned to evolocumab (140 mg every 2 weeks or 420 mg monthly) or placebo, and at 48 weeks, evolocumab reduced LDL-C by 59% compared to placebo, from a median baseline value of 92 mg/dL to 30 mg/dL. With a median follow-up of 2.2 years, evolocumab significantly reduced the primary endpoint, which was a composite of CV death, MI, stroke, hospitalization for unstable angina, or coronary revascularization, by 15% (9.8% vs. 11.3%; HR 0.85; 95% CI 0.79–0.92; $p < 0.001$), and the key secondary endpoint, which was a composite of CV death, MI, or stroke, by 20% (5.9% vs. 7.4%; HR 0.80; 95% CI 0.73–0.88; $p < 0.001$) (Table 1). Although the magnitude of the risk reduction in the primary and key secondary endpoints appeared to grow over time, there were no significant differences in CV death (HR 1.05; 95% CI 0.88–1.25) and all-cause mortality (HR 1.04; 95% CI 0.91–1.19).

Diabetes was present at baseline in 11,031 (40%) patients, and in a prespecified secondary analysis [20], similar efficacy in the primary and key secondary endpoints was observed in patients with and without diabetes. However, since patients with diabetes had a higher baseline risk, they seemed to have a greater absolute risk reduction in the primary endpoint over 3 years ((2.7%; number needed to treat 37) vs. (1.6%; number needed to treat 62); $p = 0.60$ for interaction). This benefit was driven largely by a greater absolute risk reduction in coronary revascularization, and there was no difference in the absolute risk reduction for the key secondary endpoint.

3.2.3. Cardiovascular Outcomes Trial: ODYSSEY Outcomes

ODYSSEY Outcomes [19] included 18,924 patients who had been hospitalized for an ACS 1 to 12 months prior to randomization. After a run-in period of 2 to 16 weeks on high-intensity statins, patients with LDL-C $\geq$ 70 mg/dL, non-HDL-C $\geq$ 100 mg/dL, or apolipoprotein B $\geq$ 80 mg/dL were randomized to alirocumab (75 mg every 2 weeks) or placebo. A target LDL-C level of 25 to 50 mg/dL was specified, with up-titration of alirocumab to 150 mg every 2 weeks in patients with LDL-C $\geq$ 50 mg/dL and a blinded switch to placebo in patients who consistently had LDL-C < 15 mg/dL.

In the on-treatment analysis, which excluded LDL-C values after premature treatment discontinuation or blinded switch to placebo, alirocumab reduced LDL-C by 61% from a mean LDL-C of 96.4 mg/dL to 42.3 mg/dL at 1 year, and by 54.7% from a mean LDL-C of 101.4 mg/dL to 53.3 mg/dL at 4 years. With a median follow-up of 2.8 years, alirocumab significantly reduced the primary endpoint, which was a composite of CHD death, MI, ischemic stroke, or unstable angina requiring hospitalization, by 15% (9.5% vs. 11.1%; HR 0.85; 95% CI 0.78–0.93; $p = 0.0003$), and the secondary composite endpoint of all-cause death, MI, or ischemic stroke by 14% (10.3% vs. 11.9%; HR 0.86; 95% CI 0.79–0.93; $p = 0.0003$) (Table 1). Although all-cause death was significantly lower with alirocumab, there were no significant differences in CHD death and CV death.

In a prespecified secondary analysis stratified by baseline LDL-C, patients with LDL-C $\geq$ 100 mg/dL appeared to gain the most benefit, with reductions in the primary and secondary endpoints, although the p value for interaction was not significant. Roughly 30% of patients had diabetes and considering that diabetes is associated with higher mortality after an ACS [33], a greater absolute risk reduction might have been seen in this population, and we await the results of further analysis.

3.2.4. Safety of Monoclonal Antibodies and Low LDL-C

Cholesterol is an essential component of all cell membranes and is critical to the maintenance of normal cell functions, such as gonadal hormones, adrenal function, and brain function. Therefore, theoretical concerns have been raised regarding the extremely low level of LDL-C achieved with PCSK9 mAbs [34]. Both evolocumab and alirocumab were safe and well tolerated in FOURIER [18] and ODYSSEY Outcomes [19], although longer follow-up periods are needed since both trials had a relatively short follow-up period of less than 3 years. A prespecified secondary analysis of FOURIER [35] showed a highly significant monotonic relationship between achieved LDL-C and major CV outcomes, without a significant association with safety outcomes, which is in accordance with the results of a secondary analysis from IMPROVE-IT [22]. Currently, evolocumab has been studied for up to 4 years in the open-label OSLER-1 extension study and has shown a good safety profile [36]. An open-label extension study of FOURIER is ongoing, which will investigate 6600 subjects for 5 years and will provide more information regarding its long-term safety (NCT03080935, NCT02867813). Key adverse events are briefly discussed below.

3.2.5. Muscle-Related Events

In FOURIER [18], rates of muscle-related events were similar between evolocumab and placebo (5.0% vs. 4.8%, respectively). Details of ODYSSEY Outcomes [19] have not been reported yet, but in ODYSSEY LONG TERM [32], alirocumab had a higher rate of myalgia compared to placebo (5.4% vs. 2.9%, respectively).

3.2.6. Injection Site Reactions

In FOURIER [18] and ODYSSEY Outcomes [19], injection site reactions were rare, but were significantly more frequent with both evolocumab (2.1% vs. 1.6%) and alirocumab (3.8% vs. 2.1%) compared to placebo.

3.2.7. Antidrug Antibodies

The development of bococizumab, a humanized mAb with approximately 3% of the murine sequence remaining, was discontinued in part due to the development of a high rate of antidrug antibodies which diminished the magnitude and durability of LDL-C reduction [37]. In contrast, evolocumab and alirocumab are fully humanized mAbs, and in FOURIER [18], only 0.3% of patients developed new antidrug antibodies, and development of neutralizing antibodies did not occur in any patient. In ODYSSEY Outcomes [19], neutralizing antibodies developed in 0.4% and 0.1% of patients in the alirocumab and placebo group, respectively, and slight attenuation of LDL-C lowering over time was observed in the trial. Further analyses are needed to elucidate whether neutralizing antibodies had a negative effect or if it was mainly due to the trial design with a specified down-titration algorithm at low LDL-C levels. In a previous report of 4747 patients from 10 trials of alirocumab, neutralizing antibodies were observed in 1.3% of patients, but reductions in LDL-C were maintained over time regardless of neutralizing antibody status [38].

3.2.8. Neurocognitive Events

In FOURIER [18] and ODYSSEY Outcomes [19], there were no significant differences in neurocognitive events for both evolocumab (1.6% vs. 1.5%) and alirocumab (1.5% vs. 1.8%) compared to placebo. Cognitive function was prospectively assessed in a subgroup of patients from FOURIER in EBBINGHAUS (Evaluating PCSK9 Binding Antibody Influence on Cognitive Health in High Cardiovascular Risk Subjects) [39] using the Cambridge Neuropsychological Test Automated Battery (CANTAB), a computerized cognitive assessment tool that uses touch-screen neuropsychological tests of cognition that are specifically designed to assess central nervous system disorders and cognitive function. A total of 1204 patients were followed for a median of 19 months, and there were no significant differences in the CANTAB score between patients who received evolocumab and placebo. The ongoing 5-year extension of FOURIER includes CANTAB assessments in approximately 500 patients who had also participated in EBBINGHAUS and will provide longer-term data regarding cognition. A clinical trial of alirocumab is also ongoing, with prospective CANTAB assessments in 2200 patients with a follow-up period of 2 years (NCT02957682). Lastly, a recently reported mendelian randomization study provides reassurance, as genetic variants in PCSK9 and 3-hydroxy-3-methylglutaryl-coenzyme A reductase (HMGCR, the target of statins) showed no causal effects of low LDL-C on the risk of Alzheimer's disease, Parkinson's disease, and dementia [40].

3.2.9. New-Onset Diabetes

Meta-analyses of randomized trials have shown a dose-dependent relationship between statins and risk of incident diabetes, with a higher risk in patients receiving intensive-dose therapy compared with moderate-dose therapy (OR 1.12; 95% CI 1.04–1.22) [41,42]. Mendelian randomization studies with genetic variants in PCSK9 have also shown an increased risk of diabetes [43,44], and whether PCSK9 mAbs carry a risk of development of diabetes has been a matter of concern. In a prespecified secondary analysis of FOURIER [20], evolocumab did not increase the risk of new-onset diabetes in patients without diabetes at baseline (HR 1.05; 95% CI 0.94–1.17), including in those with prediabetes (HR 1.00; 95% CI 0.89–1.13). Levels of HbA1c and fasting plasma glucose were similar between the evolocumab and placebo groups over time in patients with diabetes, prediabetes, or normoglycemia. In ODYSSEY Outcomes [19], alirocumab did not increase the risk of new-onset diabetes in patients without diabetes at baseline (9.6% vs. 10.1%) and did not have an adverse effect of worsening diabetes or diabetic complications in patients with diabetes at baseline (18.8% vs. 21.2%).

3.2.10. Cost-Effectiveness

The efficacy and safety of evolocumab and alirocumab in FOURIER [18] and ODYSSEY Outcomes [19] are very promising, but with a hefty price tag of $14,000 per year in the United States. Three cost-effectiveness analyses [45–47] incorporating data from FOURIER have been reported, with an incremental cost-effectiveness ratio (ICER) of $268,600 to $450,000 per quality-adjusted life-year (QALY), and a 60% to 70% reduction from current prices would be needed to achieve a societally acceptable ICER of $100,000 per QALY [48].

3.2.11. Intracellular PCSK9 Inhibitors: Inclisiran

Inclisiran is a long-acting, subcutaneously delivered siRNA targeting PCSK9 messenger RNA (mRNA) [49]. It is attached to an N-acetylgalactosamine moiety, which facilitates selective uptake into liver cells via the asialoglycoprotein receptor [50]. After binding intracellularly to the RNA-induced silencing complex (RISC), it specifically cleaves mRNA encoding PCSK9 [49].

In ORION-1 [51], a phase 2 study of inclisiran, 501 patients at high risk for CVD, of which 118 (24%) patients had diabetes at baseline, with elevated LDL-C despite maximum tolerated dose of statins were randomized to receive a single dose of placebo or 200, 300, or 500 mg inclisiran or two doses (on days 1 and 90) of placebo or 100, 200, or 300 mg inclisiran. Inclisiran reduced PCSK9 and LDL-C levels in a dose-dependent manner, and at 6 months, LDL-C reductions of 27.9% to 41.9% after a single dose and 35.5% to 52.6% after two doses ($p < 0.001$ for all comparisons vs. placebo) were observed. The two-dose 300 mg inclisiran regimen produced the greatest reduction in LDL-C with a mean reduction of 64.2 mg/dL from baseline, with 48% and 66% of patients achieving LDL-C <50 mg/dL and <70 mg/dL, respectively. There seemed to be no adverse events related to inclisiran, but 5% of patients who received inclisiran experienced injection-site reactions whereas no injection-site reactions occurred in patients assigned to placebo.

Whether there is a clinically meaningful difference in intracellular and extracellular PCSK9 inhibition remains unknown. However, inclisiran has the advantage of a twice-yearly injection and a lower manufacturing cost when compared with PCSK9 mAbs, which require an injection every 2 to 4 weeks with a substantial cost burden [52]. ORION-10 (NCT03399370) and ORION-11 (NCT03400800), phase 3 studies recruiting a combined total of 3,000 patients with ASCVD and elevated LDL-C despite maximum tolerated dose of statins, are ongoing, and ORION-4 [53], a randomized CV outcome trial of inclisiran in 15,000 patients with stable ASCVD and LDL-C $\geq$ 100 mg/dL with a median follow-up of 5 years, is planned.

3.3. Bempedoic Acid

Bempedoic acid (ETC-1002) is a once-daily, orally administered prodrug that inhibits adenosine triphosphate citrate lyase (ACL), a key enzyme upstream of HMGCR involved in the synthesis of fatty acids and cholesterol [54,55]. It also activates 5′-adenosine monophosphate-activated protein kinase (AMPK), reducing the activity of acetyl-CoA carboxylase (ACC) and HMGCR, the rate-limiting enzymes of fatty acid and cholesterol synthesis, respectively [55]. The activation of AMPK also targets phosphoenolpyruvate carboxykinase and glucose-6-phosphatase, enzymes with a crucial role in gluconeogenesis and liver glucose production [55] and seems to have a favorable effect on glucose regulation in animal models [56]. In the liver, the prodrug is activated by very long-chain acyl-CoA synthetase-1 (ASCVL1), but since skeletal muscle does not express ASCVL1, it remains in its inactive form and can potentially avoid the myotoxicity associated with statins [54].

In phase 2 clinical trials, bempedoic acid has shown significant LDL-C reductions of up to 50% when combined with ezetimibe [57], and in patients with type 2 diabetes [56], it was associated with a nonsignificant reduction in fasting and postprandial glucose concentrations, along with a nonsignificant tendency of improved glycemic control in a 24-h continuous glucose monitoring assessment compared to placebo. A phase 3 CV outcome trial involving 12,600 high-risk patients who are statin intolerant is ongoing (CLEAR Outcomes; NCT02993406).

3.4. Fibrates

Fibrates are agonists of peroxisome proliferator-activated receptor alpha (PPARα), mediating transcription factors that control lipoprotein metabolism [58]. They improve the lipid profile of diabetic dyslipidemia by decreasing TG level and increasing HDL-C level, but recent trials have failed to show a benefit in outcomes, both with monotherapy [59] and in addition to a statin [60]. In the ACCORD (Action to Control Cardiovascular Risk in Diabetes) lipid trial [60], the addition of fenofibrate to simvastatin in high-risk patients with T2DM did not reduce the primary endpoint, which was a composite of MI, stroke, or CV death, with a mean follow-up of 4.7 years. However, in a prespecified subgroup analysis, a possible benefit for patients with both a high baseline TG level $\geq$ 204 mg/dL and a low baseline level of HDL-C $\leq$ 34 mg/dL was suggested, with similar post hoc subgroup analysis in the FIELD (Fenofibrate Intervention and Event Lowering in Diabetes) study [61]. PROMINENT (Pemafibrate to Reduce Cardiovascular OutcoMes by Reducing Triglycerides IN patiENts with diabeTes; NCT03071692), a CV outcome trial of pemafibrate is currently underway, and will investigate 10,000 patients with T2DM who have TG level $\geq$ 200 mg/dL and HDL-C level $\leq$ 40 mg/dL despite concomitant statin therapy.

3.5. Omega 3 Fatty Acids

Omega 3 fatty acids (eicosapentaenoic acid (EPA) and docosahexaenoic acid (DHA)) decrease TG level, but they have produced even more inconsistent results than fibrates [62]. We await the results of two ongoing CV outcome trials, REDUCE-IT (Reduction of Cardiovascular Events with Icosapent Ethyl-Intervention Trial; NCT01492361) and STRENGTH (A Long-Term Outcomes Study to Assess STatin Residual Risk Reduction With EpaNova in HiGh Cardiovascular Risk PatienTs With Hypertriglyceridemia; NCT02104817), which will evaluate the effect of omega 3 fatty acids on top of statins in high-risk patients with mixed dyslipidemia.

4. Therapeutic Strategies for Dyslipidemia in Patients with Type 2 Diabetes

Some guidelines recommend an LDL-C treatment target, while others recommend a specific statin intensity without an LDL-C target (Table 2) [2,9–12]. However, they all agree on statins as the first-line treatment, and problems arise when patients cannot achieve their target LDL-C level or are statin-intolerant. First and foremost, accurate identification of true statin intolerance is of vital importance, since many patients are able to tolerate statins when rechallenged. Statin-associated muscle symptoms are usually not of pharmacological origin, but rather a consequence of patient perceptions that statins can cause muscle symptoms, combined with the high background prevalence of muscle symptoms in middle-aged and elderly patients [14]. In GAUSS-3 (Goal Achievement After Utilizing an Anti-PCSK9 Antibody in Statin Intolerant Subjects 3) [63], 511 patients intolerant to multiple statins were rechallenged in a double-blinded manner with one half of the patients randomized to atorvastatin 20 mg and the other half randomized to placebo for the first 10 weeks, with subsequent crossover to the alternate treatment group. During this rechallenge phase, 26.5% of patients experienced muscle symptoms with placebo but not with atorvastatin, supporting the aforementioned notion, although true statin intolerance clearly exists in some patients as 42.6% of patients experienced muscle symptoms with atorvastatin but not with placebo.

Ezetimibe should be used in patients who fail to achieve their target LDL-C level with statins alone or are statin-intolerant in view of the benefits proven in IMPROVE-IT [17]. We prefer ezetimibe over PCSK9 mAbs due to its oral administration and low cost, and PCSK9 mAbs should be reserved for very high-risk patients with clinical ASCVD and LDL-C above their target level despite concomitant use of statins and ezetimibe. Gemfibrozil, a type of fibrate, inhibits statin glucuronidation and leads to the elevation of plasma statin concentration. Therefore, it should not be combined with statins because of an increased risk of myotoxicity [64]. In contrast, fenofibrate does not have a significant effect on statin glucuronidation, and there was no increased risk in the ACCORD lipid trial where fenofibrate–simvastatin combination therapy was investigated [60]. The addition of fenofibrate may be considered in high-risk patients with elevated TG level and low HDL-C level despite statin treatment, taking into account the suggested benefit in the subgroup analyses of ACCORD lipid [60] and FIELD [61].

Table 2. Recommendations for treatment of dyslipidemia in patients with T2DM.

(a) Guidelines with LDL-C treatment target recommendations			
	ASCVD (+)	**ASCVD (−) and Other CVD Risk Factors (+)**	**ASCVD (−) and Other CVD Risk Factors (−)**	**Nonstatin Treatment**
2017 AACE/ACE [11]	LDL-C < 55 mg/dL	LDL-C < 70 mg/dL	LDL-C < 100 mg/dL	If LDL-C above target, consider ezetimibe. If ASCVD (+), also consider PCSK9 mAb.
	ASCVD (+)	**ASCVD (−) and Other CVD Risk Factors (+)**	**ASCVD (−) and Other CVD Risk Factors (−)**	**Nonstatin Treatment**
2016 ESC/EAS [10] (2017 update on PCSK9 mAb [65])	LDL-C < 70 mg/dL	LDL-C < 70 mg/dL (age > 40 years)	LDL-C < 100 mg/dL	If LDL-C above target, consider ezetimibe. If LDL-C > 100 mg/dL in ASCVD (+), also consider PCSK9 mAb. *
	CHD (+)		**CHD (−)**	**Nonstatin Treatment**
2017 JAS [12]	LDL-C <100 mg/dL (LDL-C <70 mg/dL in very high-risk patients)		LDL-C < 120 mg/dL (age 40–74 years)	If LDL-C above target, consider combination therapy (no specific drug indicated).

(b) Guidelines with statin intensity recommendations			
	ASCVD (+) or LDL-C ≥ 190 mg/dL	**ASCVD (−) and LDL-C 70–189 mg/dL**	**Nonstatin Treatment**
2013 ACC/AHA [9] (2017 update on nonstatin treatment [66])	High-intensity statin (moderate-intensity statin if age > 75 years)	Moderate-intensity statin (age 40–75 years)	If <50% LDL-C reduction, consider ezetimibe. If ASCVD (+), also consider PCSK9 mAb.
	ASCVD (+)	**ASCVD (−)**	**Nonstatin Treatment**
2018 ADA [2]	High-intensity statin	Moderate-intensity statin (age ≥ 40 years)	If LDL-C ≥ 70 mg/dL in ASCVD (+), consider ezetimibe or PCSK9 mAb.

AACE/ACE = American Association of Clinical Endocrinologists/American College of Endocrinology; ACC/AHA = American College of Cardiology/American Heart Association; ADA = American Diabetes Association; ASCVD = atherosclerotic cardiovascular disease; CAD = coronary artery disease; CVD = cardiovascular disease; ESC/EAS = European Society of Cardiology/European Atherosclerosis Society; JAS = Japan Atherosclerosis Society; LDL-C = low-density lipoprotein cholesterol, mAb = monoclonal antibody; PCSK9 = proprotein convertase subtilisin/kexin type 9; * LDL-C threshold, the starting value on which treatment decisions for a PCSK9 mAb are based, which is different from the LDL-C goal, the aim of therapeutic intervention.

5. Conclusions

Statins have been and will remain the cornerstone of treatment of dyslipidemia in patients with diabetes. However, residual risk remains, and nonstatin treatment with ezetimibe and PCSK9 mAbs has an evolving role with proven benefits on CV outcomes. As the most potent LDL-C-lowering agent available on the market, PCSK9 mAbs have huge expectations, but their long-term safety remains to be established, and their prices must come down for them to be cost-effective. Inclisiran would be a huge addition to our arsenal especially if their market price is much cheaper than PCSK9 mAbs. The potential of bempedoic acid to avoid the myotoxicity of statins with a favorable effect on glucose metabolism is exciting, and we eagerly await the results of CV outcome trials of fibrates and omega 3 fatty acids to further understand the role of high TG level in CVD. We have made huge progress but much remains to be done, and further advances to improve the care of our patients should be anticipated.

Author Contributions: K.S. contributed to data interpretation and wrote the first draft of the manuscript. Y.S. conceptualized the study. All authors reviewed the manuscript and approved the final version.

Funding: The authors received no specific funding for this work. Article Processing Charge was sponsored by MDPI.

Acknowledgments: The authors thank Wendy Gray, self-employed, for English editing.

Conflicts of Interest: The authors declare no conflict of interest.

References

1. Ryden, L.; Grant, P.J.; Anker, S.D.; Berne, C.; Cosentino, F.; Danchin, N.; Deaton, C.; Escaned, J.; Hammes, H.P.; Huikuri, H.; et al. ESC Guidelines on diabetes, pre-diabetes, and cardiovascular diseases developed in collaboration with the EASD: The Task Force on diabetes, pre-diabetes, and cardiovascular diseases of the European Society of Cardiology (ESC) and developed in collaboration with the European Association for the Study of Diabetes (EASD). *Eur. Heart J.* **2013**, *34*, 3035–3087. [CrossRef] [PubMed]

2. American Diabetes Association. 9. Cardiovascular Disease and Risk Management: Standards of Medical Care in Diabetes-2018. *Diabetes Care* **2018**, *41*, S86–S104. [CrossRef]

3. Haffner, S.M.; Lehto, S.; Ronnemaa, T.; Pyorala, K.; Laakso, M. Mortality from coronary heart disease in subjects with type 2 diabetes and in nondiabetic subjects with and without prior myocardial infarction. *N. Engl. J. Med.* **1998**, *339*, 229–234. [CrossRef] [PubMed]

4. Juutilainen, A.; Lehto, S.; Ronnemaa, T.; Pyorala, K.; Laakso, M. Type 2 diabetes as a "coronary heart disease equivalent": An 18-year prospective population-based study in Finnish subjects. *Diabetes Care* **2005**, *28*, 2901–2907. [CrossRef] [PubMed]

5. Verges, B. Pathophysiology of diabetic dyslipidaemia: Where are we? *Diabetologia* **2015**, *58*, 886–899. [CrossRef] [PubMed]

6. Colhoun, H.M.; Betteridge, D.J.; Durrington, P.N.; Hitman, G.A.; Neil, H.A.; Livingstone, S.J.; Thomason, M.J.; Mackness, M.I.; Charlton-Menys, V.; Fuller, J.H. Primary prevention of cardiovascular disease with atorvastatin in type 2 diabetes in the Collaborative Atorvastatin Diabetes Study (CARDS): Multicentre randomised placebo-controlled trial. *Lancet* **2004**, *364*, 685–696. [CrossRef]

7. Collins, R.; Armitage, J.; Parish, S.; Sleigh, P.; Peto, R. MRC/BHF Heart Protection Study of cholesterol-lowering with simvastatin in 5963 people with diabetes: A randomised placebo-controlled trial. *Lancet* **2003**, *361*, 2005–2016. [PubMed]

8. Kearney, P.M.; Blackwell, L.; Collins, R.; Keech, A.; Simes, J.; Peto, R.; Armitage, J.; Baigent, C. Efficacy of cholesterol-lowering therapy in 18,686 people with diabetes in 14 randomised trials of statins: A meta-analysis. *Lancet* **2008**, *371*, 117–125. [CrossRef] [PubMed]

9. Stone, N.J.; Robinson, J.G.; Lichtenstein, A.H.; Bairey Merz, C.N.; Blum, C.B.; Eckel, R.H.; Goldberg, A.C.; Gordon, D.; Levy, D.; Lloyd-Jones, D.M.; et al. 2013 ACC/AHA guideline on the treatment of blood cholesterol to reduce atherosclerotic cardiovascular risk in adults: A report of the American College of Cardiology/American Heart Association Task Force on Practice Guidelines. *J. Am. Coll. Cardiol.* **2014**, *63*, 2889–2934. [CrossRef] [PubMed]

10. Catapano, A.L.; Graham, I.; De Backer, G.; Wiklund, O.; Chapman, M.J.; Drexel, H.; Hoes, A.W.; Jennings, C.S.; Landmesser, U.; Pedersen, T.R.; et al. 2016 ESC/EAS Guidelines for the Management of Dyslipidaemias. *Eur. Heart J.* **2016**, *37*, 2999–3058. [CrossRef] [PubMed]

11. Jellinger, P.S.; Handelsman, Y.; Rosenblit, P.D.; Bloomgarden, Z.T.; Fonseca, V.A.; Garber, A.J.; Grunberger, G.; Guerin, C.K.; Bell, D.S.H.; Mechanick, J.I.; et al. American Association of Clinical Endocrinologists and American College of Endocrinology Guidelines for Management of Dyslipidemia and Prevention of Cardiovascular Disease. *Endocr. Pract.* **2017**, *23*, 1–87. [CrossRef] [PubMed]

12. Japan Atherosclerosis Society. *Japan Atherosclerosis Society (JAS) Guidelines for Prevention of Atherosclerotic Cardiovascular Diseases*, 2017 ed.; Japan Atherosclerosis Society: Tokyo, Japan, 2017.

13. Boekholdt, S.M.; Hovingh, G.K.; Mora, S.; Arsenault, B.J.; Amarenco, P.; Pedersen, T.R.; LaRosa, J.C.; Waters, D.D.; DeMicco, D.A.; Simes, R.J.; et al. Very low levels of atherogenic lipoproteins and the risk for cardiovascular events: A meta-analysis of statin trials. *J. Am. Coll. Cardiol.* **2014**, *64*, 485–494. [CrossRef] [PubMed]

14. Newman, C.B.; Tobert, J.A. Statin intolerance: Reconciling clinical trials and clinical experience. *JAMA* **2015**, *313*, 1011–1012. [CrossRef] [PubMed]

15. Altmann, S.W.; Davis, H.R., Jr.; Zhu, L.J.; Yao, X.; Hoos, L.M.; Tetzloff, G.; Iyer, S.P.; Maguire, M.; Golovko, A.; Zeng, M.; et al. Niemann-Pick C1 Like 1 protein is critical for intestinal cholesterol absorption. *Science* **2004**, *303*, 1201–1204. [CrossRef] [PubMed]

16. Temel, R.E.; Tang, W.; Ma, Y.; Rudel, L.L.; Willingham, M.C.; Ioannou, Y.A.; Davies, J.P.; Nilsson, L.M.; Yu, L. Hepatic Niemann-Pick C1-like 1 regulates biliary cholesterol concentration and is a target of ezetimibe. *J. Clin. Investig.* **2007**, *117*, 1968–1978. [CrossRef] [PubMed]

17. Cannon, C.P.; Blazing, M.A.; Giugliano, R.P.; McCagg, A.; White, J.A.; Theroux, P.; Darius, H.; Lewis, B.S.; Ophuis, T.O.; Jukema, J.W.; et al. Ezetimibe Added to Statin Therapy after Acute Coronary Syndromes. *N. Engl. J. Med.* **2015**, *372*, 2387–2397. [CrossRef] [PubMed]

18. Sabatine, M.S.; Giugliano, R.P.; Keech, A.C.; Honarpour, N.; Wiviott, S.D.; Murphy, S.A.; Kuder, J.F.; Wang, H.; Liu, T.; Wasserman, S.M.; et al. Evolocumab and Clinical Outcomes in Patients with Cardiovascular Disease. *N. Engl. J. Med.* **2017**, *376*, 1713–1722. [CrossRef] [PubMed]

19. Steg, P.G. Evaluation of Cardiovascular Outcomes after an Acute Coronary Syndrome during Treatment with Alirocumab—ODYSSEY OUTCOMES. In Proceedings of the American College of Cardiology Annual Scientific Session (ACC 2018), Orlando, FL, USA, 10–12 March 2018.

20. Sabatine, M.S.; Leiter, L.A.; Wiviott, S.D.; Giugliano, R.P.; Deedwania, P.; De Ferrari, G.M.; Murphy, S.A.; Kuder, J.F.; Gouni-Berthold, I.; Lewis, B.S.; et al. Cardiovascular safety and efficacy of the PCSK9 inhibitor evolocumab in patients with and without diabetes and the effect of evolocumab on glycaemia and risk of new-onset diabetes: A prespecified analysis of the FOURIER randomised controlled trial. *Lancet Diabetes Endocrinol.* **2017**, *5*, 941–950. [CrossRef]

21. Giugliano, R.P.; Cannon, C.P.; Blazing, M.A.; Nicolau, J.C.; Corbalan, R.; Spinar, J.; Park, J.G.; White, J.A.; Bohula, E.; Braunwald, E. Benefit of Adding Ezetimibe to Statin Therapy on Cardiovascular Outcomes and Safety in Patients With vs. Without Diabetes: Results from IMPROVE-IT. *Circulation* **2017**. [CrossRef]

22. Giugliano, R.P.; Wiviott, S.D.; Blazing, M.A.; De Ferrari, G.M.; Park, J.G.; Murphy, S.A.; White, J.A.; Tershakovec, A.M.; Cannon, C.P.; Braunwald, E. Long-term Safety and Efficacy of Achieving Very Low Levels of Low-Density Lipoprotein Cholesterol: A Prespecified Analysis of the IMPROVE-IT Trial. *JAMA Cardiol.* **2017**, *2*, 547–555. [CrossRef] [PubMed]

23. Abifadel, M.; Varret, M.; Rabes, J.P.; Allard, D.; Ouguerram, K.; Devillers, M.; Cruaud, C.; Benjannet, S.; Wickham, L.; Erlich, D.; et al. Mutations in PCSK9 cause autosomal dominant hypercholesterolemia. *Nat. Genet.* **2003**, *34*, 154–156. [CrossRef] [PubMed]

24. Cohen, J.; Pertsemlidis, A.; Kotowski, I.K.; Graham, R.; Garcia, C.K.; Hobbs, H.H. Low LDL cholesterol in individuals of African descent resulting from frequent nonsense mutations in PCSK9. *Nat. Genet.* **2005**, *37*, 161–165. [CrossRef] [PubMed]

25. Cohen, J.C.; Boerwinkle, E.; Mosley, T.H., Jr.; Hobbs, H.H. Sequence variations in PCSK9, low LDL, and protection against coronary heart disease. *N. Engl. J. Med.* **2006**, *354*, 1264–1272. [CrossRef] [PubMed]

26. Zhao, Z.; Tuakli-Wosornu, Y.; Lagace, T.A.; Kinch, L.; Grishin, N.V.; Horton, J.D.; Cohen, J.C.; Hobbs, H.H. Molecular characterization of loss-of-function mutations in PCSK9 and identification of a compound heterozygote. *Am. J. Hum. Genet.* **2006**, *79*, 514–523. [CrossRef] [PubMed]

27. Goldstein, J.L.; Brown, M.S. A century of cholesterol and coronaries: From plaques to genes to statins. *Cell* **2015**, *161*, 161–172. [CrossRef] [PubMed]

28. Zhang, D.W.; Lagace, T.A.; Garuti, R.; Zhao, Z.; McDonald, M.; Horton, J.D.; Cohen, J.C.; Hobbs, H.H. Binding of proprotein convertase subtilisin/kexin type 9 to epidermal growth factor-like repeat A of low density lipoprotein receptor decreases receptor recycling and increases degradation. *J. Biol. Chem.* **2007**, *282*, 18602–18612. [CrossRef] [PubMed]

29. Preiss, D.; Mafham, M. PCSK9 inhibition: The dawn of a new age in cholesterol lowering? *Diabetologia* **2017**, *60*, 381–389. [CrossRef] [PubMed]

30. Karatasakis, A.; Danek, B.A.; Karacsonyi, J.; Rangan, B.V.; Roesle, M.K.; Knickelbine, T.; Miedema, M.D.; Khalili, H.; Ahmad, Z.; Abdullah, S.; et al. Effect of PCSK9 Inhibitors on Clinical Outcomes in Patients With Hypercholesterolemia: A Meta-Analysis of 35 Randomized Controlled Trials. *J. Am. Heart Assoc.* **2017**, *6*. [CrossRef] [PubMed]

31. Sabatine, M.S.; Giugliano, R.P.; Wiviott, S.D.; Raal, F.J.; Blom, D.J.; Robinson, J.; Ballantyne, C.M.; Somaratne, R.; Legg, J.; Wasserman, S.M.; et al. Efficacy and safety of evolocumab in reducing lipids and cardiovascular events. *N. Engl. J. Med.* **2015**, *372*, 1500–1509. [CrossRef] [PubMed]

32. Robinson, J.G.; Farnier, M.; Krempf, M.; Bergeron, J.; Luc, G.; Averna, M.; Stroes, E.S.; Langslet, G.; Raal, F.J.; El Shahawy, M.; et al. Efficacy and safety of alirocumab in reducing lipids and cardiovascular events. *N. Engl. J. Med.* **2015**, *372*, 1489–1499. [CrossRef] [PubMed]

33. Donahoe, S.M.; Stewart, G.C.; McCabe, C.H.; Mohanavelu, S.; Murphy, S.A.; Cannon, C.P.; Antman, E.M. Diabetes and mortality following acute coronary syndromes. *JAMA* **2007**, *298*, 765–775. [CrossRef] [PubMed]

34. Olsson, A.G.; Angelin, B.; Assmann, G.; Binder, C.J.; Bjorkhem, I.; Cedazo-Minguez, A.; Cohen, J.; von Eckardstein, A.; Farinaro, E.; Muller-Wieland, D.; et al. Can LDL cholesterol be too low? Possible risks of extremely low levels. *J. Intern. Med.* **2017**, *281*, 534–553. [CrossRef] [PubMed]

35. Giugliano, R.P.; Pedersen, T.R.; Park, J.G.; De Ferrari, G.M.; Gaciong, Z.A.; Ceska, R.; Toth, K.; Gouni-Berthold, I.; Lopez-Miranda, J.; Schiele, F.; et al. Clinical efficacy and safety of achieving very low LDL-cholesterol concentrations with the PCSK9 inhibitor evolocumab: A prespecified secondary analysis of the FOURIER trial. *Lancet* **2017**, *390*, 1962–1971. [CrossRef]

36. Koren, M.J.; Sabatine, M.S.; Giugliano, R.P.; Langslet, G.; Wiviott, S.D.; Kassahun, H.; Ruzza, A.; Ma, Y.; Somaratne, R.; Raal, F.J. Long-term Low-Density Lipoprotein Cholesterol-Lowering Efficacy, Persistence, and Safety of Evolocumab in Treatment of Hypercholesterolemia: Results up to 4 years from the Open-Label OSLER-1 Extension Study. *JAMA Cardiol.* **2017**, *2*, 598–607. [CrossRef] [PubMed]

37. Ridker, P.M.; Tardif, J.C.; Amarenco, P.; Duggan, W.; Glynn, R.J.; Jukema, J.W.; Kastelein, J.J.P.; Kim, A.M.; Koenig, W.; Nissen, S.; et al. Lipid-Reduction Variability and Antidrug-Antibody Formation with Bococizumab. *N. Engl. J. Med.* **2017**, *376*, 1517–1526. [CrossRef] [PubMed]

38. Roth, E.M.; Goldberg, A.C.; Catapano, A.L.; Torri, A.; Yancopoulos, G.D.; Stahl, N.; Brunet, A.; Lecorps, G.; Colhoun, H.M. Antidrug Antibodies in Patients Treated with Alirocumab. *N. Engl. J. Med.* **2017**, *376*, 1589–1590. [CrossRef] [PubMed]

39. Giugliano, R.P.; Mach, F.; Zavitz, K.; Kurtz, C.; Im, K.; Kanevsky, E.; Schneider, J.; Wang, H.; Keech, A.; Pedersen, T.R.; et al. Cognitive Function in a Randomized Trial of Evolocumab. *N. Engl. J. Med.* **2017**, *377*, 633–643. [CrossRef] [PubMed]

40. Benn, M.; Nordestgaard, B.G.; Frikke-Schmidt, R.; Tybjaerg-Hansen, A. Low LDL cholesterol, PCSK9 and HMGCR genetic variation, and risk of Alzheimer's disease and Parkinson's disease: Mendelian randomisation study. *BMJ* **2017**, *357*, j1648. [CrossRef] [PubMed]

41. Sattar, N.; Preiss, D.; Murray, H.M.; Welsh, P.; Buckley, B.M.; de Craen, A.J.; Seshasai, S.R.; McMurray, J.J.; Freeman, D.J.; Jukema, J.W.; et al. Statins and risk of incident diabetes: A collaborative meta-analysis of randomised statin trials. *Lancet* **2010**, *375*, 735–742. [CrossRef]

42. Preiss, D.; Seshasai, S.R.; Welsh, P.; Murphy, S.A.; Ho, J.E.; Waters, D.D.; DeMicco, D.A.; Barter, P.; Cannon, C.P.; Sabatine, M.S.; et al. Risk of incident diabetes with intensive-dose compared with moderate-dose statin therapy: A meta-analysis. *JAMA* **2011**, *305*, 2556–2564. [CrossRef] [PubMed]

43. Schmidt, A.F.; Swerdlow, D.I.; Holmes, M.V.; Patel, R.S.; Fairhurst-Hunter, Z.; Lyall, D.M.; Hartwig, F.P.; Horta, B.L.; Hypponen, E.; Power, C.; et al. PCSK9 genetic variants and risk of type 2 diabetes: A mendelian randomisation study. *Lancet Diabetes Endocrinol.* **2017**, *5*, 97–105. [CrossRef]

44. Ference, B.A.; Robinson, J.G.; Brook, R.D.; Catapano, A.L.; Chapman, M.J.; Neff, D.R.; Voros, S.; Giugliano, R.P.; Davey Smith, G.; Fazio, S.; et al. Variation in PCSK9 and HMGCR and Risk of Cardiovascular Disease and Diabetes. *N. Engl. J. Med.* **2016**, *375*, 2144–2153. [CrossRef] [PubMed]

45. Kazi, D.S.; Penko, J.; Coxson, P.G.; Moran, A.E.; Ollendorf, D.A.; Tice, J.A.; Bibbins-Domingo, K. Updated Cost-effectiveness Analysis of PCSK9 Inhibitors Based on the Results of the FOURIER Trial. *JAMA* **2017**, *318*, 748–750. [CrossRef] [PubMed]

46. Fonarow, G.C.; Keech, A.C.; Pedersen, T.R.; Giugliano, R.P.; Sever, P.S.; Lindgren, P.; van Hout, B.; Villa, G.; Qian, Y.; Somaratne, R.; et al. Cost-effectiveness of Evolocumab Therapy for Reducing Cardiovascular Events in Patients With Atherosclerotic Cardiovascular Disease. *JAMA Cardiol.* **2017**, *2*, 1069–1078. [CrossRef] [PubMed]

47. Arrieta, A.; Hong, J.C.; Khera, R.; Virani, S.S.; Krumholz, H.M.; Nasir, K. Updated Cost-effectiveness Assessments of PCSK9 Inhibitors From the Perspectives of the Health System and Private Payers: Insights Derived From the FOURIER Trial. *JAMA Cardiol.* **2017**, *2*, 1369–1374. [CrossRef] [PubMed]

48. Bonow, R.O.; Harrington, R.A.; Yancy, C.W. Cost-effectiveness of PCSK9 Inhibitors: Proof in the Modeling. *JAMA Cardiol.* **2017**, *2*, 1298–1299. [CrossRef] [PubMed]

49. Fitzgerald, K.; White, S.; Borodovsky, A.; Bettencourt, B.R.; Strahs, A.; Clausen, V.; Wijngaard, P.; Horton, J.D.; Taubel, J.; Brooks, A.; et al. A Highly Durable RNAi Therapeutic Inhibitor of PCSK9. *N. Engl. J. Med.* **2017**, *376*, 41–51. [CrossRef] [PubMed]

50. Nordestgaard, B.G.; Nicholls, S.J.; Langsted, A.; Ray, K.K.; Tybjaerg-Hansen, A. Advances in lipid-lowering therapy through gene-silencing technologies. *Nat. Rev. Cardiol.* **2018**. [CrossRef] [PubMed]

51. Ray, K.K.; Landmesser, U.; Leiter, L.A.; Kallend, D.; Dufour, R.; Karakas, M.; Hall, T.; Troquay, R.P.; Turner, T.; Visseren, F.L.; et al. Inclisiran in Patients at High Cardiovascular Risk with Elevated LDL Cholesterol. *N. Engl. J. Med.* **2017**, *376*, 1430–1440. [CrossRef] [PubMed]

52. Khvorova, A. Oligonucleotide Therapeutics—A New Class of Cholesterol-Lowering Drugs. *N. Engl. J. Med.* **2017**, *376*, 4–7. [CrossRef] [PubMed]

53. Inclisiran Marches on: ORION-4 Cardiovascular Outcomes Study Launches. Available online: https://www.pcsk9forum.org/inclisiran-marches-orion-4-cardiovascular-outcomes-study-launches/ (accessed on 1 April 2018).

54. Pinkosky, S.L.; Newton, R.S.; Day, E.A.; Ford, R.J.; Lhotak, S.; Austin, R.C.; Birch, C.M.; Smith, B.K.; Filippov, S.; Groot, P.H.; et al. Liver-specific ATP-citrate lyase inhibition by bempedoic acid decreases LDL-C and attenuates atherosclerosis. *Nat. Commun.* **2016**, *7*, 13457. [CrossRef] [PubMed]

55. Lemus, H.N.; Mendivil, C.O. Adenosine triphosphate citrate lyase: Emerging target in the treatment of dyslipidemia. *J. Clin. Lipidol.* **2015**, *9*, 384–389. [CrossRef] [PubMed]

56. Gutierrez, M.J.; Rosenberg, N.L.; Macdougall, D.E.; Hanselman, J.C.; Margulies, J.R.; Strange, P.; Milad, M.A.; McBride, S.J.; Newton, R.S. Efficacy and safety of ETC-1002, a novel investigational low-density lipoprotein-cholesterol-lowering therapy for the treatment of patients with hypercholesterolemia and type 2 diabetes mellitus. *Arterioscler. Thromb. Vasc. Biol.* **2014**, *34*, 676–683. [CrossRef] [PubMed]

57. Thompson, P.D.; MacDougall, D.E.; Newton, R.S.; Margulies, J.R.; Hanselman, J.C.; Orloff, D.G.; McKenney, J.M.; Ballantyne, C.M. Treatment with ETC-1002 alone and in combination with ezetimibe lowers LDL cholesterol in hypercholesterolemic patients with or without statin intolerance. *J. Clin. Lipidol.* **2016**, *10*, 556–567. [CrossRef] [PubMed]

58. Staels, B.; Dallongeville, J.; Auwerx, J.; Schoonjans, K.; Leitersdorf, E.; Fruchart, J.C. Mechanism of action of fibrates on lipid and lipoprotein metabolism. *Circulation* **1998**, *98*, 2088–2093. [CrossRef] [PubMed]

59. Keech, A.; Simes, R.J.; Barter, P.; Best, J.; Scott, R.; Taskinen, M.R.; Forder, P.; Pillai, A.; Davis, T.; Glasziou, P.; et al. Effects of long-term fenofibrate therapy on cardiovascular events in 9795 people with type 2 diabetes mellitus (the FIELD study): Randomised controlled trial. *Lancet* **2005**, *366*, 1849–1861. [CrossRef]

60. Ginsberg, H.N.; Elam, M.B.; Lovato, L.C.; Crouse, J.R., 3rd; Leiter, L.A.; Linz, P.; Friedewald, W.T.; Buse, J.B.; Gerstein, H.C.; Probstfield, J.; et al. Effects of combination lipid therapy in type 2 diabetes mellitus. *N. Engl. J. Med.* **2010**, *362*, 1563–1574. [CrossRef] [PubMed]

61. Scott, R.; O'Brien, R.; Fulcher, G.; Pardy, C.; D'Emden, M.; Tse, D.; Taskinen, M.R.; Ehnholm, C.; Keech, A. Effects of fenofibrate treatment on cardiovascular disease risk in 9795 individuals with type 2 diabetes and various components of the metabolic syndrome: The Fenofibrate Intervention and Event Lowering in Diabetes (FIELD) study. *Diabetes Care* **2009**, *32*, 493–498. [CrossRef] [PubMed]

62. Reiner, Z. Hypertriglyceridaemia and risk of coronary artery disease. *Nat. Rev. Cardiol.* **2017**, *14*, 401–411. [CrossRef] [PubMed]

63. Nissen, S.E.; Stroes, E.; Dent-Acosta, R.E.; Rosenson, R.S.; Lehman, S.J.; Sattar, N.; Preiss, D.; Bruckert, E.; Ceska, R.; Lepor, N.; et al. Efficacy and Tolerability of Evolocumab vs. Ezetimibe in Patients With Muscle-Related Statin Intolerance: The GAUSS-3 Randomized Clinical Trial. *JAMA* **2016**, *315*, 1580–1590. [CrossRef] [PubMed]

64. Franssen, R.; Vergeer, M.; Stroes, E.S.; Kastelein, J.J. Combination statin-fibrate therapy: Safety aspects. *Diabetes Obes. Metab.* **2009**, *11*, 89–94. [CrossRef] [PubMed]

65. Landmesser, U.; Chapman, M.J.; Stock, J.K.; Amarenco, P.; Belch, J.J.F.; Boren, J.; Farnier, M.; Ference, B.A.; Gielen, S.; Graham, I.; et al. 2017 Update of ESC/EAS Task Force on practical clinical guidance for proprotein convertase subtilisin/kexin type 9 inhibition in patients with atherosclerotic cardiovascular disease or in familial hypercholesterolaemia. *Eur. Heart J.* **2017**. [CrossRef] [PubMed]

66. Lloyd-Jones, D.M.; Morris, P.B.; Ballantyne, C.M.; Birtcher, K.K.; Daly, D.D., Jr.; DePalma, S.M.; Minissian, M.B.; Orringer, C.E.; Smith, S.C., Jr. 2017 Focused Update of the 2016 ACC Expert Consensus Decision Pathway on the Role of Non-Statin Therapies for LDL-Cholesterol Lowering in the Management of Atherosclerotic Cardiovascular Disease Risk: A Report of the American College of Cardiology Task Force on Expert Consensus Decision Pathways. *J. Am. Coll. Cardiol.* **2017**, *70*, 1785–1822. [CrossRef] [PubMed]

 *diseases*

Review

Lipid Target in Very High-Risk Cardiovascular Patients: Lesson from PCSK9 Monoclonal Antibodies

Giovanni Ciccarelli, Saverio D'Elia, Michele De Paulis, Paolo Golino and Giovanni Cimmino *

Department of Cardio-Thoracic and Respiratory Sciences, Section of Cardiology, University of Campania "Luigi Vanvitelli", 80131 Naples, Italy; ciccarelli.giovanni@gmail.com (G.C.); saveriodelia85@gmail.com (S.D.); mdepaulis89@gmail.com (M.D.P.); paolo.golino@unicampania.it (P.G.)
* Correspondence: giovanni.cimmino@unicampania.it; Tel.: +39-081-7064175; Fax: +39-081-7064234

Received: 14 January 2018; Accepted: 14 March 2018; Published: 17 March 2018

Abstract: The role of low-density lipoproteins (LDLs) as a major risk factor for cardiovascular disease has been demonstrated by several epidemiological studies. The molecular basis for LDLs in atherosclerotic plaque formation and progression is not completely unraveled yet. Pharmacological modulation of plasma LDL-C concentrations and randomized clinical trials addressing the impact of lipid-lowering interventions on cardiovascular outcome have clearly shown that reducing plasma LDL-C concentrations results in a significant decrease in major cardiovascular events. For many years, statins have represented the most powerful pharmacological agents available to lower plasma LDL-C concentrations. In clinical trials, it has been shown that the greater the reduction in plasma LDL-C concentrations, the lower the rate of major cardiovascular events, especially in high-risk patients, because of multiple risk factors and recurrent events. However, in a substantial number of patients, the recommended LDL target is difficult to achieve because of different factors: genetic background (familial hypercholesterolemia), side effects (statin intolerance), or high baseline plasma LDL-C concentrations. In the last decade, our understanding of the molecular mechanisms involved in LDL metabolism has progressed significantly and the key role of proprotein convertase subtilisin/kexin type 9 (PCSK9) has emerged. This protein is an enzyme able to bind the LDL receptors (LDL-R) on hepatocytes, favoring their degradation. Blocking PCSK9 represents an intriguing new therapeutic approach to decrease plasma LDL-C concentrations, which in recent studies has been demonstrated to also result in a significant reduction in major cardiovascular events.

Keywords: lipoproteins; atherosclerosis; cardiovascular risk; statin; PCSK9

---

1. Introduction

A strong correlation between lipid plasma levels and atherosclerotic cardiovascular disease (ASCVD) has been clearly shown over the years [1]. Lipid deposition, mainly low-density lipoprotein (LDL), occurs in the arterial wall of almost all vascular districts, defining the atherosclerotic process, with focal clinical manifestations based on the organ damaged: heart (coronary artery disease), brain (cerebrovascular disease), and/or limbs (peripheral vascular disease) [2]. Plaque complication results in exposure of prothrombotic material to the flowing blood, leading to acute thrombus formation, which may result in an acute medical emergency, such as acute coronary syndrome (ACS), stroke, or related problems [3–5]. Lowering plasma LDL concentrations is highly recommended for patients with hyperlipidemia and multiple risk factors, such as diabetes, hypertension, smoking status, chronic kidney disease, or peripheral artery disease [6,7]. According to the current guidelines, diet and weight loss represent the first "medical" approach [1,8]. However, pharmacological modulation of plasma lipid concentrations is quite often necessary according to the target lipid levels.

The majority of clinical trials have shown the efficacy of statins in reducing major cardiovascular events (MACE) in both primary and secondary prevention [9,10]. Moreover, because of their low cost, they are sustainable from the economic point of view. However, despite the proven efficacy, in a subset of patients achievement of the plasma cholesterol target by statins remains an unmet clinical need because of: (1) intolerance (defined as muscle pain and/or liver dysfunction) [11]; (2) not powerful enough to achieve the target [12]; (3) unfavorable genetic background, (i.e., heterozygous familial hypercholesterolemia (FH) or homozygous FH [13]).

In the last 15 years, new insight into the basic mechanisms involved in cholesterol metabolism has been gained, and new pharmacological targets have been identified [14]. The proprotein convertase subtilisin/kexin type 9 (PCSK9) is an example of rapid pre-clinical and clinical progression between their discovery in 2003 and the present because of the potential the innovative therapeutic scenario opened up [15]. This protein is a key player in the clearance of LDL particles and its inhibition seems to be highly effective in reducing plasma LDL-C concentrations [16]. This review, starting with the role of cholesterol in the pathogenesis of atherosclerosis in high-risk patients, will discuss the novelty of the PCSK9 approach and the available data on the safety and effectiveness of its inhibition in reducing plasma LDL-C concentrations.

2. Traditional Pharmacological Approaches for Dyslipidaemias: From Bench to Bedside

2.1. Atherosclerotic High-Risk Patients

Atherosclerosis is a multifactorial, chronic "inflammatory-degenerative" disease of the arterial tree [3,17], recognizing in the atheroma its pathological substrate [18]. Conventional (hypertension, diabetes, smoking, hyperlipidemia, age, sex) and less conventional (impaired glucose tolerance; impaired fasting glucose, apolipoprotein B (ApoB); apolipoprotein A-I (ApoAI), triglycerides; triglyceride-rich lipoproteins[TGRLs], small and dense LDL, oxidized-LDL, antibodies against oxidized-LDL, Lipoprotein (a), homocysteine, high-sensitivity C-reactive protein) cardiovascular risk factors are the main determinants for the impairment of the endothelial protective properties [19]. The resulting endothelial dysfunction [20–23] with deposition of circulating LDL-C in the arterial wall is the initial step of the atherosclerotic process [24]. Within the subendothelial space, LDL-C become more susceptible to oxidation due to local reactive oxygen species (ROS) production, triggering the atherosclerotic cascade that leads to plaque formation, progression, and destabilization [2]. In each step of this process, the role of inflammation and immunity is now well defined [3,25].

Patients with diabetes mellitus (DM) are considered at high cardiovascular risk, with a ~10-fold increased risk in their lifetime [26–28]. Increased plaque vulnerability [29], advanced glycosylation end products (AGEs) formation [30,31], enzyme-mediated endothelial damage [30], and a modified lipid profile (with an increase in VLDLs and "small and dense" LDLs) [32] indicates that diabetic patients may have accelerated atherosclerosis.

Patients with chronic kidney disease (CKD) also show an increased risk for ASCVD [33], mainly related to a loss of renal parenchyma, which accelerates atherosclerosis [34], a chronic inflammation status with elevated CRP levels, and reduced renal clearance of several cytokines, such as IL-1β, IL-6, and TNF-α [35], as well as raised levels of angiotensin II and parathormone, which contribute to increased ROS production and calcium deposition in the vessel wall, thus inducing endothelial dysfunction [33]. Coronary plaques from CKD patients show extensive calcification and increased presence of thrombotic events [36].

Autopsy studies have shown that fibrous plaques are more common in the femoral arteries [37,38]. Peripheral atherosclerotic disease occurs in the context of multiple disease processes that interfere with exercise ability. Potential mechanisms include the conventional risk factors in a local setting of reduced blood flow, altered muscle metabolism, and impaired angiogenesis, leading to limb discomfort and functional limitation [39].

2.2. Major Drugs Used in LDL Cholesterol-Lowering Strategies

Several randomized clinical trials have unequivocally shown that lipid-lowering therapies are associated with MACE reduction [40–42]. Despite dietary approach, physical activity, and weight loss being the strategies recommended first [43], many pharmacological strategies have been identified to modulate plasma lipid levels.

(a) *Statins:* Inhibition of HMG-CoA reductase by statins has been the first therapeutic approach. These drugs block the endogenous synthesis of cholesterol in the liver, resulting in a reduction of intracellular cholesterol. This causes induction of LDL-R expression of the hepatocyte surface, which in turn leads to enhanced clearance of LDL particles from the blood. Several trials have highlighted an improvement in terms of cardiovascular morbidity and mortality as well as the need for coronary artery interventions using statins [44,45]. This protective effect is of greater magnitude (a) if statin treatment is started earlier and lasts a longer time; (b) if a high dose is given [9,10,46,47]; and/or (c) if the percentage LDL-C reduction from baseline value is higher [48]. The PROVE IT-TIMI 22 trial (Pravastatin or Atorvastatin Evaluation and Infection Therapy–Thrombolysis in Myocardial Infarction 22) demonstrated that plasma LDL-C concentrations at baseline are an important predictor of the benefit of intensive lipid-lowering therapy [48]. Indeed, decreasing LDL-cholesterol baseline levels is an additional benefit of intensive treatment with statins compared with moderate-dose therapy declines. Consequently, current guidelines suggest that HMG-CoA reductase inhibitors represent the first choice for patients with hypercholesterolemia or combined hyperlipidemia [1].

(b) *Selective cholesterol absorption inhibitors:* By inhibiting intestinal cholesterol and phytosterol absorption protein (Niemann-Pick C1-Like 1, NPC1L1) present on jejunal intestinal cells, ezetimibe decreases intestinal cholesterol absorption from dietary sources and from bile, resulting in a reduction in hepatic cholesterol concentration and circulating LDL-C by up to 20% [14]. The IMPROVE-IT trial (Improved Reduction of Outcomes: Vytorin Efficacy International Trial) concluded that the combination of ezetimibe plus simvastatin as compared to simvastatin alone in patients with acute coronary syndromes resulted in a further lowering of plasma LDL-C concentrations by up to 50 mg/dL, with an associated improvement in cardiovascular outcomes [49].

(c) *Fibrates:* Fibric acid derivatives, or fibrates, are agonists of α isoform of peroxisome proliferator-activated receptor (PPAR). Activation of this receptor may result in several modifications of the plasma lipid profile [50]. Five major effects have been characterized by the use of fibrates: (1) Lipoprotein lipolysis induction via lipoprotein lipase activity and apoC-III inhibition; (2) increased hepatic fatty acid uptake and reduction of hepatic triglyceride production (these two effects result in hypotriglyceridemic action); (3) formation of LDL particles with a higher affinity for the own receptor, thus resulting in a higher rate of LDL particles removal; (4) reduction in cholesteryl ester and triglycerides exchange between VLDL and HDL, leading to decreased plasma levels of triglyceride-rich lipoproteins; and (5) increase in the production of apoA-I and apoA-II in liver, thus inducing a higher HDL production and promoting reverse cholesterol transport [51,52]. However, because of the variable results from the clinical trials investigating the impact of fibrates on clinical outcomes, either in primary and secondary prevention, and problems linked to safety, their role remains limited to selected patients with diabetes, metabolic syndrome, or dyslipidemia [53–56].

A large amount of clinical trial data indicate that the central point of all lipid-lowering strategies for high-risk patients is the percentage change from baseline rather than any predefined target. This view is corroborated by the following findings: (1) on-treatment LDL-C plasma concentrations levels do not predict CVD risk rates, whereas baseline LDL-C plasma concentrations do; (2) the correlation between LDL reduction and CVD risk reduction within each study is at best curvilinear even in high-risk, high-cholesterol populations where the average on-treatment LDL is still significantly distant from the predefined target of 100 mg/dL; (3) despite the same percentage of LDL reduction, more benefits have been reported in subjects with higher baseline LDL (and whose on-treatment LDL stays higher than 100 mg/dL). Despite the achieved LDL-C goal, in very high-risk patients recurrent events might occur. By taking into account the linear correlation between LDL-C level reduction and the risk of fatal or nonfatal MACE, a further percentage variation from their baseline (even if it is low) might result in additional benefits [57]. PCSK9 inhibition by monoclonal antibodies opens up a new way to achieve this goal.

3. PCSK9 Inhibition: A Route to Very Low LDL-C Plasma Concentrations

Because of a residual cardiovascular risk in patients not at the target LDL level or with a percentage variation from baseline still low, researchers are still looking for the best pharmacological strategy.

The discovery of proprotein convertase subtilisin/kexin type 9 (PCSK9) has created a new frontier for better management of dyslipidemia. Indeed, in 2003 Abifadel et al., studying gene mutations responsible for familial hypercholesterolemia (FH), found that mutations in the *PCSK9* gene cause dominant hypercholesterolemia in pedigree analysis. Interestingly, a gain of function (GOF) due to a missense mutation in this gene was the cause of the disease [58]. To date, three gene mutations are known to cause FH: (1) the LDL receptor itself (FH1); (2) apolipoprotein (apo) B (FH2), a ligand of LDL receptors; and (3) a GOF mutation in PCSK9 (FH3).

A better understanding of the PCSK9 pathway showed that this molecule promotes the degradation of LDL receptors by forming an enzyme–substrate complex, mainly in the liver. The cell- surface complex LDL-R/LDL is transported to the endosomes via endocytosis, where LDLs are released in acid conditions. Within the endosome, LDL is degraded to amino acids and cholesterol while LDL-R is transported back to the cell surface, but, if bound to PCSK9, LDL-R will be degraded too (Figure 1). Therefore, PCSK9 inhibition may be an effective strategy to promote LDL-R recycling, thus reducing circulating LDL particles (Figure 1). The rate of plasma LDL-C decrease is approximately 45–60%, whether used alone or in combination with a statin. Addition of an anti-PCSK9 antibody to standard therapy—with statin alone, or statin combined with Ezetimibe—resulted in a further reduction of plasma LDL-C concentrations (up to 60%) and a halved cardiovascular event rate compared to the placebo [59].

Evolocumab and Alirocumab are "fully" humanized anti-PCSK9 antibodies, while Bococizumab is a humanized monoclonal antibody. These three monoclonal antibodies are currently under investigation in extensive clinical programs and trials, namely PROFICIO, ODYSSEY, and SPIRE. The latest was discontinued in November 2016.

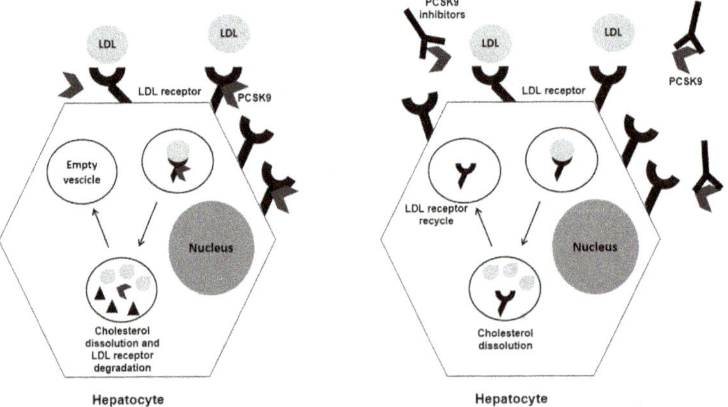

Figure 1. Schematic view of PCSK9 activity and effects of its inhibition.

As reported above, the PROFICIO (Program to Reduce LDL-C and Cardiovascular Outcomes Following Inhibition of PCSK9 in Different Populations) is the development program for Evolocumab, a monoclonal antibody already approved by the FDA and EMA. Several trials of this program have already reported efficacy (with a mean of LDL-C reduction up to 57%), safety, and durable effects in different populations [14,60,61], even patients affected by LDL receptor abnormalities such as homozygous or heterozygous Familial Hypercholesterolemia (FH) [62,63]. The first direct effect of PCSK9 inhibition by evolocumab on atherosclerotic plaque comes from the GLAGOV study, published in 2016 [64]. Coronary atherosclerotic lesions have been evaluated by intravascular ultrasound in patients receiving evolocumab or placebo on top of statin therapy. Among patients with angiographic evidence of coronary artery disease and on chronic statin therapy, the PCSK9 inhibitor evolocumab resulted in a greater change in atheroma volume (−0.95% vs. +0.05% of the placebo group) and a greater proportion of patients with plaque regression with a mean plasma LDL-C concentrations in the active drug group of 36.6 mg/dL [64]. Moreover, the latest results available from the FOURIER study, a large-scale outcome study in 27,564 ASCVD patients on statin therapy presented in 2017, showed that, with the addition of evolocumab, the mean plasma LDL-C concentrations dropped to 30 mg/dL. In addition, in a median follow-up period of 2.2 years, there was a decrease in the rate of the composite primary endpoint of cardiovascular death, myocardial infarction, stroke, hospitalization due to unstable angina, and coronary artery revascularization (percutaneous coronary intervention, coronary artery bypass graft) of 15% compared with the placebo [65].

Another big program is the ODYSSEY trials testing Alirocumab. As for evolocumab, several data from this program have already been published, showing the power and safety of this approach in reducing LDL-C plasma concentrations by up to 47% in patients at high risk [14,60,66]. The latest results from the ODYSSEY Outcomes trial, presented at the American College of Cardiology in 2018, indicate that treatment with alirocumab reduced cardiovascular outcomes and all-cause deaths by 15% in a group of high-risk patients. These data reinforce the previous data from the FOURIER trial and expand them because of a longer outcome and a higher-risk population. A mean of 53.3 mg/dL of LDL-C plasma concentration was achieved in the alirocumab group, with a percentage variation of 54.7. The benefits were more marked in those patients with the highest LDL cholesterol at baseline. In the pre-specified post hoc analysis by LDL-C level at baseline, patients with an LDL-C ≥100 mg/dL experienced reductions in all endpoints. In the alirocumab group a 24% reduction in MACE was reported with an absolute risk reduction (ARR) of 3.4%. In detail, CHD death was reduced by 28% (ARR 0.9%), CV death by 31% (ARR 1.3%), and all-cause death by 28% (ARR 1.7%).

The FOURIER and the ODYSSEY Outcomes trials consistently show benefits from PCSK9 inhibitors, not only in terms of preventing nonfatal events such as heart attacks but in actually preserving life. Thus, the future is now.

A major warning raised by some recent data exploring the impact of PCSK9 inhibitors in cardiovascular outcome, was the possible higher rate of neurocognitive adverse events [61]. However, more recent analysis seems to not confirm this correlation. In two pre-specified analyses of the FOURIER study, despite slightly more injection-site reactions in the evolocumab arm, the clinical efficacy and safety of the PCSK9 therapy have been confirmed [67], with no difference between the groups especially regarding cognitive impairment (EBBINGHAUSS study) [68] and new-onset diabetes [69].

Moreover, the beneficial effects of lipid-lowering therapy with a non-statin agent added to high- or moderate-intensity statin therapy have also been reported in patients with symptomatic lower extremity peripheral artery disease (PAD), including those without prior MI or stroke [70].

Finally, a more recent meta-analysis found evidence of a significantly greater reduction of plasma LDL-C concentrations in patients treated with evolocumab than those treated with alirocumab [66].

The third monoclonal PCSK9 inhibitor antibody, namely Bococizumab, was discontinued in November 2016 because of neutralizing antibodies [71]. However, its effects have been described in two large-scale cardiovascular outcomes trials, SPIRE-1 and SPIRE-2, where a total of 27,438 participants with either a history of cardiovascular disease or familial hypercholesterolemia or high risk for cardiovascular disease were randomized to either bococizumab subcutaneously or a matching placebo. Despite the discontinuation, the results from both SPIRE trials confirmed the efficacy of PCSK9 inhibition in terms of very low plasma LDL-C concentrations and cardiovascular event reduction [72].

At the time of this review, new pharmacological approaches are under development to safely inhibit PCSK9 (Table 1). The latest strategy involves the use of a small interfering RNA (siRNA) [73].

The siRNA molecules, namely inclisiran, engage the natural pathway of RNA interference (RNAi) by binding intracellularly to the RNA-induced silencing complex (RISC), enabling it to cleave messenger RNA (mRNA) molecules specifically encoding PCSK9. The cleaved mRNA is degraded and thus unavailable for protein translation, which results in decreased levels of the PCSK9 protein [74].

Table 1. PCSK9 under development.

Drugs	Type	Status	Study
Evolocumab	Monoclonal Ab	Approved	Proficio Program
Alirocumab	Monoclonal Ab	Approved	Odyssey Program
Bococizumab	Monoclonal Ab	Discontinued	Spire Program
Inclisiran	Silent RNA	On approval	Orion 1
LGT-209	Monoclonal Ab	Discontinued	-
RG7652	Monoclonal Ab	Phase 2	Equator
ALN-PC	RNAinhibitor	Phase 1 ev/Preclinical sc	-
Adnectin BMS-962476	modified binding protein	Phase 1	-
EGF-A peptide	synthetic peptide	Preclinical	-

Based on the current available evidence, a PCSK9 inhibitor should be considered in specific subsets of patients:

- Patients with ASCVD at very high risk of an adverse prognosis, with persistent elevated plasma LDL-C concentrations despite maximally tolerated statin alone or in combination with ezetimibe therapy.
- Patients with ASCVD at very high risk with persistent elevated plasma LDL-C concentrations, who show intolerance to the appropriate doses of at least three statins.
- Familial hypercholesterolemia patients without clinically diagnosed ASCVD, at high cardiovascular risk, with persistent elevated plasma LDL-C concentrations despite maximally tolerated statin plus ezetimibe therapy [75].

4. Metabolic Effects of PCSK9 Inhibition: Beyond LDL-C Reduction

In humans, PCSK9 expression has been reported in different organs (i.e., brain, kidney, pancreas, liver, and small intestine) [76] and cells (i.e., endothelium, smooth muscle cells, and macrophages) [77,78], suggesting systemic and local homeostatic effects. Based on these data, blocking PCSK9 may result in unexpected effects.

4.1. PCSK9 and Glucose Metabolism

It is known that perturbation of cholesterol metabolism may expose patients to increased risk of diabetes, since statin treatment [79] and genetic variant of HMGCoA reductase [80] have been associated with a higher prevalence of type 2 diabetes. Moreover, statin use and genetic inhibition of HMGCoA reductase are associated with increased body weight [80–82], thus resulting in metabolic syndrome and diabetes predisposition. These data seem to be confirmed by the evidence that patients with familial hypercholesterolemia showed a lower prevalence of glucose metabolism impairment [83].

Inhibition of PCSK9 in pancreatic cells decreases cholesterol accumulation, which results in glucose metabolism impairment and reduction of insulin secretion, thus favoring diabetes status [84]. However, contradictory data are available on this issue, since PCSK9 loss of function (LOF) is not associated with increased risk of diabetes [85] while the PCSK9 R46L variant seems to be associated with insulin resistance [86]. From a metabolic point of view, PCSK9, by decreasing LDL-R activity, reduces cholesterol concentrations within the pancreatic beta-cell, finally resulting in increased beta-cell function and insulin secretion. Moreover, the direct effect of PCSK9 on pancreatic delta-cell leads to increased insulin secretion, thus resulting in beneficial effects on carbohydrate homeostasis. On the other hand, PCSK9 induces insulin-regulated secretion via SREBP-1C with increased glucose, HbA1c, and HOMA-IR index, leading to a detrimental effect on carbohydrate homeostasis, thus counterbalancing the beneficial effect of the other metabolic pathway. Based on these data, PCSK9 inhibition should result in a neutral effect [84].

The latest analysis of the published clinical trials, looking specifically at the new onset of diabetes, seems to support this neutral effect [67,69]. However, the relatively short time of observation of these studies may represent a major limitation. More recently, Cao et al. [87] published a systematic review and meta-analysis of a total of 18 studies including 26,123 participants treated with alirocumab or evolocumab and without diabetes. PCSK9 inhibition had no significant impact on new-onset diabetes mellitus and glucose homeostasis, regardless of PCSK9-mAb type, participant characteristics, treatment duration, treatment method, and differences in control treatment, thus adding new evidence for the neutral effect of PCSK9 on glycemic control.

4.2. PCSK9 and Lipogenesis

PCSK9 not only targets LDL-R but may regulate other steps and key players in lipogenesis [76]. Recently, Verbeek et al. [88] have analyzed the lipid profile and lipoprotein subfractions by nuclear magnetic resonance in carriers of R46L variant that are responsible for a PCSK9 LOF. The population enrolled was part of the European Prospective Investigation into Cancer and Nutrition (EPIC) study, designed to investigate the relationships between diet, nutritional status, lifestyle, environmental factors, and the incidence of cancer and other chronic diseases. The data from this analysis confirm and expand previous results [89] on the effects on other lipoproteins (such as Apolipoprotein B (ApoB), Lp (a), secretory phospholipase A2, and Lp-PLA2(beyond the already known decrease in LDL-C plasma concentrations. Specifically, IDL and VLDL particles were subject to a variation of −18% and −16%, respectively. No effect on HDL-C plasma concentrations has been reported [88]. These observations may indicate that the positive effects of PCSK9 inhibition, especially on patients at high risk, may be related, at least in part, to the reduction of most atherogenic lipoproteins, lifelong exposure to which is associated with an increased cardiovascular risk. It has been reported that PCSK9 circulates in the flowing blood as lipoprotein-bound, mainly associated with ApoB within LDL and VLDL particles, thus suggesting an involvement in another molecular pathway beyond the LDL. Confounding data are available at the time of this article on the interaction between PCSK9 and VLDL-R, ApoE-R2, and CD36. Because these receptors are involved in the basal and post-prandial plasma triglycerides levels, PCSK9 may be indirectly involved in this pathway [90]. A gene silencing study with siRNA and clinical observations in carriers of GOF and LOF variants have been performed.

Compared to patients carrying an LDL-R mutation, individuals with a PCSK9 GOF variant show higher levels of VLDL, IDL, and triglycerides beyond the increased LDL-C plasma concentrations. On the other hand, individuals with a PCSK9 LOF variant, especially the R46L, show a reduction in all ApoB-containing lipoproteins [90], and attenuated levels of fasting and postprandial triglycerides [91]. Based on these observations, a putative role of PCSK9 in the clearance of triglyceride-rich lipoproteins (TGRLs) may be postulated. Similar results have been obtained by inhibiting PCSK9 via gene silencing or viral transfection, thus suggesting a non-direct effect of PCSK9 on triglyceride metabolism [90]. Moreover, a recent study confirms that Alirocumab reduces ApoB levels in both IDL and LDL, most likely due to increased LDL degradation [92]. It is known that ApoB turnover is strongly influenced by the uptake of lipoproteins containing ApoB, mediated by LDL-R [93]. Results from the FOURIER trials indicate that PCSK9 inhibition is associated with a mild–moderate reduction in triglyceride levels [59]. The biological mechanism underlying this effect may include several pathways beyond the increased LDL-R activity that are associated with the increased catabolism of TGRLs [94]. It is known that PCSK9 modulates lipoprotein assembly and secretion by the intestine and the liver and affects TGRL and fatty acid uptake in peripheral tissues via expression of the very-low-density lipoprotein receptor (VLDL-R), the ApoE2 receptor, and the CD36 receptor [16,76]. The receptor for VLDL modulates the extra-hepatic metabolism of TGRLs in concert with lipoprotein lipase, thus contributing to the delivery of fatty acids to these peripheral tissues, especially in the heart, skeletal muscle, and adipose tissue, where it is highly expressed.

However, despite the reduction in triglycerides (up to 17.3%), ApoB (up to 56%) and the positive outcomes in the clinical trials testing PCSK9 inhibition by both monoclonal antibodies, alirocumab and evolocumab, the clinical impact of the non-LDL cholesterol concentrations on CV risk remain unclear.

Summarizing, the lower plasma triglyceride concentrations and the decreased postprandial lipemia may be the results of the following mechanisms [94]: (a) reduced intestine ApoB48 production, leading to decreased chylomicron secretion; (b) increased ApoB degradation, resulting in reduced ApoB-rich lipoproteins; (c) increased chylomicrons, chylomicron remnants, and VLDL remnants clearance via increased LDL-related protein 1 activity and CD36 scavenger receptor upregulation; (d) VLDL receptors and ApoE receptors.

4.3. PCSK9 and HDL Particles

Finally, the available clinical trials (OSLER [95], ODYSSEY Long Term [96], and FOURIER [65]) indicate that PCSK9 inhibition may also affect HDL metabolism. A modest increase in plasma HDL-C and apoA1 concentrations (estimated at less than 10%) has been reported. This effect may be the result of two potential mechanisms: (a) the reduced number of LDL particles leads to a decreased transfer of cholesterol from HDL to LDL particles [97,98]; (b) the blockage of PCSK9 reduces cholesterol ester transfer protein (CETP) activity, thus decreasing heteroexchange of lipids between TGRLs and HDL particles [99].

5. "Very Low Is Better": End of Story?

The progression of pharmacological modulation of lipid metabolism and the available data on safety and efficacy of the current strategies indicate that more ambitious targets in terms of lipid lowering can be achieved. Randomized clinical trials published to date clearly showed that PCSK9 inhibitors may result in plasma LDL-C concentrations lower than 50 mg/dL easily with no safety concern [100]. More interestingly, a recent report from Ference et al. quantified the reduction of cardiovascular events in about 20% per decrease of 1.00 mmol/L (39 mg/dL) in plasma LDL-C concentrations [101]. These effects mediated by PCSK9 inhibitors are independent and additive with statins. Thus, taking into account the fact that inhibition of PCSK9 results in a strong reduction of plasma LDL-C concentrations, the benefit expected will be higher, especially in patients with an elevated baseline value.

Based on these results, the latest guidelines from the American Association of Clinical Endocrinologists and American College of Endocrinology [102] identify an "extreme risk" group (Table 2a), beyond the very high-risk group already defined in the European Society of Cardiology Guidelines 2016 [103] (Table 2b), characterized by progressive atherosclerotic cardiovascular disease, including unstable angina that persists after achieving an LDL-C less than 70 mg/dL, or established clinical ASCVD with diabetes, stage 3 or 4 CKD, and/or HeFH, or in those with a history of premature ASCVD (<55 years of age for males or <65 years of age for females) in which a LDL-C goal of less than 55 mg/dL is recommended. The goal of 30 mg/dL, as indicated in most of the PCSK9 inhibitors clinical trials, is desirable.

However, how safe it is to go "very low" remains to be elucidated [104] and will be a matter for the long-term outcomes study to determine in the next few years. On this safety concern, some epidemiological studies available to date indicate a different association between cholesterol plasma concentration and neoplastic risk, hemorrhagic stroke, depression and anxiety, and nervous and immune system dysfunction. Specifically, decreased plasma LDL-C concentrations have been associated with increased cancer risk [105]. In a study from Benn et al. including participants from the Copenhagen City Heart Study and the Copenhagen General Population Study, low plasma LDL-C concentrations were firmly associated with increased cancer risk, but genetically decreased LDL-C was not [106]. Whether this association is causal remains unclear. Low LDL-C plasma concentrations may be a consequence, rather than a cause, of the neoplastic disease, thus may not cause cancer per se. Pre-specified clinical trials designed to address this safety issue are warranted.

Table 2. Current recommendations.

a. Recommendations of AACE 2017 (American Association of Clinical Endocrinologists and American College of Endocrinology)			Achieved LDL-C by PCSK9 mAb [60]
EXTREME risk: Progressive atherosclerotic cardiovascular disease, including unstable angina that persists after achieving an LDL-C less than 70 mg/dL, or established clinical ASCVD with diabetes, stage 3 or 4 CKD, and/or HeFH, or in those with a history of premature ASCVD (<55 years of age for males or <65 years of age for females)	LDL-C goal of less than 55 mg/dL is recommended	Grade A; BEL 1	**−43.8 to −55.2%** in HeHF with baseline value ≥100 mg/dL **48 mg/dL** in high ASCVD
VERY HIGH risk: Established or recent hospitalization for ACS; coronary, carotid or peripheral vascular disease; diabetes or stage 3 or 4 CKD with one or more risk factors; a calculated 10-year risk greater than 20%; or HeFH	LDL-C goal of less than 70 mg/dL is recommended	Grade A; BEL 1	**−36.3 to −54%** In patients with prior CV disease + LDL-C ≥ 70 mg/dL,
HIGH risk: An ASCVD equivalent including diabetes or stage 3 or 4 CKD with no other risk factors, or individuals with two or more risk factors and a 10-year risk of 10–20%	LDL-C goal of less than 100 mg/dL is recommended	Grade A; BEL 1	**−36.3 to −54%** In patients with CV risk factors + LDL-C ≥ 100 mg/dL
MODERATE risk: Two or fewer risk factors and a calculated 10-year risk of less than 10%	LDL-C goal of less than 100 mg/dL is recommended	Grade A; BEL 1	**−58.7%** Patients not having adequate control of their hypercholesterolemia based on their individual level of CVD risk
LOW risk: No risk factors	For individuals at low risk (i.e., with no risk factors), an LDL-C goal of less than 130 mg/dL is recommended.	Grade A; BEL 1	
b. Recommendations of ESC (European Society of Cardiology) Guidelines 2016			
VERY HIGH CV risk: -Documented CVD -DM or type-1 DM with TOG -Severe RD: GFR <30 mg/mL/1.73 m^2 −10-year risk SCORE ≥10%	**VERY-HIGH CV risk:** LDL-c goal <70 mg/dL (1.8 mmol/L) and/or 50% reduction if baseline is 70–135 mg/dL (1.8–3.5 mmol/L)	CLASS I LEVEL B	
HIGH CV risk: -Markedly elevated single risk factor −10-year risk SCORE ≥5% and <10% -Moderate RD: GFR 30–59 mg/mL/1.73 m^2	**HIGH CV risk:** LDL-c goal <100 mg/L (2.6 mmol/L) or 50% reduction if baseline is 100–200 mg/dL (2.6–5.1 mmol/L)	CLASS I LEVEL B	
MODERATE CV risk: −10-year risk SCORE ≥1% and <5%	**MODERATE CV risk:** LDL-c goal <115 mg/dL (3.0 mmol/L)	CLASS IIa LEVEL C	
LOW CV risk: −10-year risk SCORE <1%			

Author Contributions: G.Ci. and P.G. conceptualized the study. G.C., S.D. and M.D.P. wrote the manuscript. G.Ci. revised the manuscript and the final version. All authors contributed to data interpretation and approved the manuscript.

Conflicts of Interest: The authors declare no conflict of interest.

References

1. Piepoli, M.F.; Hoes, A.W.; Agewall, S.; Albus, C.; Brotons, C.; Catapano, A.L.; Cooney, M.T.; Corra, U.; Cosyns, B.; Deaton, C.; et al. 2016 European guidelines on cardiovascular disease prevention in clinical practice: The sixth joint task force of the european society of cardiology and other societies on cardiovascular disease prevention in clinical practice (constituted by representatives of 10 societies and by invited experts)developed with the special contribution of the european association for cardiovascular prevention & rehabilitation (EACPR). *Eur. Heart J.* **2016**, *37*, 2315–2381. [PubMed]

2. Cimmino, G.; Conte, S.; Morello, A.; D'Elia, S.; Marchese, V.; Golino, P. The complex puzzle underlying the pathophysiology of acute coronary syndromes: From molecular basis to clinical manifestations. *Expert Rev. Cardiovasc. Ther.* **2012**, *10*, 1533–1543. [CrossRef] [PubMed]

3. Cimmino, G.; Loffredo, F.S.; Morello, A.; D'Elia, S.; De Palma, R.; Cirillo, P.; Golino, P. Immune-inflammatory activation in acute coronary syndromes: A look into the heart of unstable coronary plaque. *Curr. Cardiol. Rev.* **2016**, *13*, 110–111. [CrossRef]

4. Cimmino, G.; Ciccarelli, G.; Golino, P. Role of tissue factor in the coagulation network. *Semin. Thromb. Hemost.* **2015**, *41*, 708–717. [CrossRef] [PubMed]

5. Cimmino, G.; D'Amico, C.; Vaccaro, V.; D'Anna, M.; Golino, P. The missing link between atherosclerosis, inflammation and thrombosis: Is it tissue factor? *Expert Rev. Cardiovasc.Ther.* **2011**, *9*, 517–523. [CrossRef] [PubMed]

6. Ray, K.K.; Cannon, C.P. Intensive statin therapy in acute coronary syndromes: Clinical benefits and vascular biology. *Curr. Opin. Lipidol.* **2004**, *15*, 637–643. [CrossRef] [PubMed]

7. Berra, K. Lipid-lowering therapy today: Treating the high-risk cardiovascular patient. *J. Cardiovasc. Nurs.* **2008**, *23*, 414–421. [CrossRef] [PubMed]

8. Hernaez, A.; Castaner, O.; Goday, A.; Ros, E.; Pinto, X.; Estruch, R.; Salas-Salvado, J.; Corella, D.; Aros, F.; Serra-Majem, L.; et al. The mediterranean diet decreases LDL atherogenicity in high cardiovascular risk individuals: A randomized controlled trial. *Mol. Nutr. Food Res.* **2017**, *61*. [CrossRef] [PubMed]

9. Ford, I.; Murray, H.; McCowan, C.; Packard, C.J. Long-term safety and efficacy of lowering low-density lipoprotein cholesterol with statin therapy: 20-year follow-up of west of scotland coronary prevention study. *Circulation* **2016**, *133*, 1073–1080. [CrossRef] [PubMed]

10. Ashen, M.D.; Foody, J.M. Evidence-based guidelines for cardiovascular risk reduction: The safety and efficacy of high-dose statin therapy. *J. Cardiovasc. Nurs.* **2009**, *24*, 429–438. [CrossRef] [PubMed]

11. Stroes, E.S.; Thompson, P.D.; Corsini, A.; Vladutiu, G.D.; Raal, F.J.; Ray, K.K.; Roden, M.; Stein, E.; Tokgozoglu, L.; Nordestgaard, B.G.; et al. Statin-associated muscle symptoms: Impact on statin therapy-european atherosclerosis society consensus panel statement on assessment, aetiology and management. *Eur. Heart J.* **2015**, *36*, 1012–1022. [CrossRef] [PubMed]

12. Sniderman, A. Targets for LDL-lowering therapy. *Curr. Opin. Lipidol.* **2009**, *20*, 282–287. [CrossRef] [PubMed]

13. Orso, E.; Ahrens, N.; Kilalic, D.; Schmitz, G. Familial hypercholesterolemia and lipoprotein(a) hyperlipidemia as independent and combined cardiovascular risk factors. *Atheroscler. Suppl.* **2009**, *10*, 74–78. [CrossRef]

14. Cimmino, G.; Loffredo, F.; Arena, G.; Golino, P. Evolving concepts in LDL-lowering strategies: Are we there? *J. Clin. Exp. Cardiol.* **2016**, *7*. [CrossRef]

15. Soutar, A.K. Unexpected roles for PCSK9 in lipid metabolism. *Curr. Opin. Lipidol.* **2011**, *22*, 192–196. [CrossRef] [PubMed]

16. Akram, O.N.; Bornier, A.; Petrides, F.; Wong, G.; Lambert, G. Beyond LDL cholesterol, a new role for PCSK9. *Arterioscler. Thromb. Vasc. Biol.* **2010**, *30*, 1279–1281. [CrossRef] [PubMed]

17. Weber, C.; Noels, H. Atherosclerosis: Current pathogenesis and therapeutic options. *Nat. Med.* **2011**, *17*, 1410–1422. [CrossRef] [PubMed]

18. Segers, D.; Weinberg, P.; Krams, R. Atherosclerosis: Cell biology and lipoproteins–shear stress and inflammation in plaque formation: New evidence. *Curr. Opin. Lipidol.* **2008**, *19*, 627–628. [CrossRef] [PubMed]

19. Fruchart, J.C.; Nierman, M.C.; Stroes, E.S.; Kastelein, J.J.; Duriez, P. New risk factors for atherosclerosis and patient risk assessment. *Circulation* **2004**, *109*. [CrossRef] [PubMed]

20. Steinberg, D. Thematic review series: The pathogenesis of atherosclerosis: An interpretive history of the cholesterol controversy, part III: Mechanistically defining the role of hyperlipidemia. *J. Lipid Res.* **2005**, *46*, 2037–2051. [CrossRef] [PubMed]

21. Steinberg, D. Thematic review series: The pathogenesis of atherosclerosis. An interpretive history of the cholesterol controversy: Part II: The early evidence linking hypercholesterolemia to coronary disease in humans. *J. Lipid Res.* **2005**, *46*, 179–190. [CrossRef] [PubMed]

22. Ghazalpour, A.; Doss, S.; Yang, X.; Aten, J.; Toomey, E.M.; Van Nas, A.; Wang, S.; Drake, T.A.; Lusis, A.J. Thematic review series: The pathogenesis of atherosclerosis. Toward a biological network for atherosclerosis. *J. Lipid Res.* **2004**, *45*, 1793–1805. [CrossRef] [PubMed]

23. Steinberg, D. Thematic review series: The pathogenesis of atherosclerosis. An interpretive history of the cholesterol controversy: Part I. *J. Lipid Res.* **2004**, *45*, 1583–1593. [CrossRef] [PubMed]

24. Badimon, J.J.; Ibanez, B.; Cimmino, G. Genesis and dynamics of atherosclerotic lesions: Implications for early detection. *Cerebrovasc. Dis.* **2009**, *27*, 38–47. [CrossRef] [PubMed]

25. Libby, P.; Hansson, G.K. Inflammation and immunity in diseases of the arterial tree: Players and layers. *Circ. Res.* **2015**, *116*, 307–311. [CrossRef] [PubMed]

26. Nathan, D.M.; Group, D.E.R. The diabetes control and complications trial/epidemiology of diabetes interventions and complications study at 30 years: Overview. *Diabetes Care* **2014**, *37*, 9–16. [CrossRef] [PubMed]

27. Lee, S.I.; Patel, M.; Jones, C.M.; Narendran, P. Cardiovascular disease and type 1 diabetes: Prevalence, prediction and management in an ageing population. *Ther. Adv. Chronic Dis.* **2015**, *6*, 347–374. [CrossRef] [PubMed]

28. Raffield, L.M.; Hsu, F.C.; Cox, A.J.; Carr, J.J.; Freedman, B.I.; Bowden, D.W. Predictors of all-cause and cardiovascular disease mortality in type 2 diabetes: Diabetes heart study. *Diabetol. Metab. Syndr.* **2015**, *7*, 58. [CrossRef] [PubMed]

29. Moreno, P.R.; Murcia, A.M.; Palacios, I.F.; Leon, M.N.; Bernardi, V.H.; Fuster, V.; Fallon, J.T. Coronary composition and macrophage infiltration in atherectomy specimens from patients with diabetes mellitus. *Circulation* **2000**, *102*, 2180–2184. [CrossRef] [PubMed]

30. Tousoulis, D.; Kampoli, A.M.; Stefanadis, C. Diabetes mellitus and vascular endothelial dysfunction: Current perspectives. *Curr. Vasc. Pharmacol.* **2012**, *10*, 19–32. [CrossRef] [PubMed]

31. Funk, S.D.; Yurdagul, A., Jr.; Orr, A.W. Hyperglycemia and endothelial dysfunction in atherosclerosis: Lessons from type 1 diabetes. *Int. J. Vasc. Med.* **2012**, *2012*, 569654. [CrossRef] [PubMed]

32. Reaven, G.M. Multiple chd risk factors in type 2 diabetes: Beyond hyperglycaemia. *Diabetes Obes. Metab.* **2002**, *4* (Suppl. 1), S13–S18. [CrossRef] [PubMed]

33. Olechnowicz-Tietz, S.; Gluba, A.; Paradowska, A.; Banach, M.; Rysz, J. The risk of atherosclerosis in patients with chronic kidney disease. *Int. Urol. Nephrol.* **2013**, *45*, 1605–1612. [CrossRef] [PubMed]

34. Kon, V.; Linton, M.F.; Fazio, S. Atherosclerosis in chronic kidney disease: The role of macrophages. *Nat. Rev. Nephrol.* **2011**, *7*, 45–54. [CrossRef] [PubMed]

35. Ghanavatian, S.; Diep, L.M.; Barany, P.; Heimburger, O.; Seeberger, A.; Stenvinkel, P.; Rohani, M.; Agewall, S. Subclinical atherosclerosis, endothelial function, and serum inflammatory markers in chronic kidney disease stages 3 to 4. *Angiology* **2014**, *65*, 443–449. [CrossRef] [PubMed]

36. Amann, K.; Tyralla, K.; Gross, M.L.; Eifert, T.; Adamczak, M.; Ritz, E. Special characteristics of atherosclerosis in chronic renal failure. *Clin. Nephrol.* **2003**, *60* (Suppl. 1), S13–S21. [PubMed]

37. Dalager, S.; Falk, E.; Kristensen, I.B.; Paaske, W.P. Plaque in superficial femoral arteries indicates generalized atherosclerosis and vulnerability to coronary death: An autopsy study. *J. Vasc. Surg.* **2008**, *47*, 296–302. [CrossRef] [PubMed]

38. Matsuo, Y.; Takumi, T.; Mathew, V.; Chung, W.Y.; Barsness, G.W.; Rihal, C.S.; Gulati, R.; McCue, E.T.; Holmes, D.R.; Eeckhout, E.; et al. Plaque characteristics and arterial remodeling in coronary and peripheral arterial systems. *Atherosclerosis* **2012**, *223*, 365–371. [CrossRef] [PubMed]

39. Hamburg, N.M.; Creager, M.A. Pathophysiology of intermittent claudication in peripheral artery disease. *Circ. J.* **2017**, *81*, 281–289. [CrossRef] [PubMed]

40. Randomised trial of cholesterol lowering in 4444 patients with coronary heart disease: The scandinavian simvastatin survival study (4s). *Lancet* **1994**, *344*, 1383–1389.

41. Shepherd, J.; Cobbe, S.M.; Ford, I.; Isles, C.G.; Lorimer, A.R.; MacFarlane, P.W.; McKillop, J.H.; Packard, C.J. Prevention of coronary heart disease with pravastatin in men with hypercholesterolemia. West of scotland coronary prevention study group. *N. Engl. J. Med.* **1995**, *333*, 1301–1307. [CrossRef] [PubMed]

42. Sacks, F.M.; Pfeffer, M.A.; Moye, L.A.; Rouleau, J.L.; Rutherford, J.D.; Cole, T.G.; Brown, L.; Warnica, J.W.; Arnold, J.M.; Wun, C.C.; et al. The effect of pravastatin on coronary events after myocardial infarction in patients with average cholesterol levels. Cholesterol and recurrent events trial investigators. *N. Engl. J. Med.* **1996**, *335*, 1001–1009. [CrossRef] [PubMed]

43. Stone, N.J.; Robinson, J.G.; Lichtenstein, A.H.; Bairey Merz, C.N.; Blum, C.B.; Eckel, R.H.; Goldberg, A.C.; Gordon, D.; Levy, D.; Lloyd-Jones, D.M.; et al. 2013 ACC/AHA guideline on the treatment of blood cholesterol to reduce atherosclerotic cardiovascular risk in adults: A report of the american college of cardiology/american heart association task force on practice guidelines. *J. Am. Coll. Cardiol.* **2014**, *63*, 2889–2934. [CrossRef] [PubMed]

44. Colhoun, H.M.; Betteridge, D.J.; Durrington, P.N.; Hitman, G.A.; Neil, H.A.; Livingstone, S.J.; Thomason, M.J.; Mackness, M.I.; Charlton-Menys, V.; Fuller, J.H.; et al. Primary prevention of cardiovascular disease with atorvastatin in type 2 diabetes in the collaborative atorvastatin diabetes study (cards): Multicentre randomised placebo-controlled trial. *Lancet* **2004**, *364*, 685–696. [CrossRef]

45. Collins, R.; Armitage, J.; Parish, S.; Sleigh, P.; Peto, R. Heart Protection Study Collaborative Group. MRC/BHF heart protection study of cholesterol-lowering with simvastatin in 5963 people with diabetes: A randomised placebo-controlled trial. *Lancet* **2003**, *361*, 2005–2016. [PubMed]

46. Rosenson, R.S. Low high-density lipoprotein cholesterol and cardiovascular disease: Risk reduction with statin therapy. *Am. Heart J.* **2006**, *151*, 556–563. [CrossRef] [PubMed]

47. Athyros, V.G.; Tziomalos, K.; Gossios, T.D.; Griva, T.; Anagnostis, P.; Kargiotis, K.; Pagourelias, E.D.; Theocharidou, E.; Karagiannis, A.; Mikhailidis, D.P.; et al. Safety and efficacy of long-term statin treatment for cardiovascular events in patients with coronary heart disease and abnormal liver tests in the greek atorvastatin and coronary heart disease evaluation (Greace) study: A post-hoc analysis. *Lancet* **2010**, *376*, 1916–1922. [CrossRef]

48. Giraldez, R.R.; Giugliano, R.P.; Mohanavelu, S.; Murphy, S.A.; McCabe, C.H.; Cannon, C.P.; Braunwald, E. Baseline low-density lipoprotein cholesterol is an important predictor of the benefit of intensive lipid-lowering therapy: A prove IT-TIMI 22 (pravastatin or atorvastatin evaluation and infection therapy-thrombolysis in myocardial infarction 22) analysis. *J. Am. Coll. Cardiol.* **2008**, *52*, 914–920. [CrossRef] [PubMed]

49. Cannon, C.P.; Blazing, M.A.; Giugliano, R.P.; McCagg, A.; White, J.A.; Theroux, P.; Darius, H.; Lewis, B.S.; Ophuis, T.O.; Jukema, J.W.; et al. Ezetimibe added to statin therapy after acute coronary syndromes. *N. Engl. J. Med.* **2015**, *372*, 2387–2397. [CrossRef] [PubMed]

50. Staels, B.; Dallongeville, J.; Auwerx, J.; Schoonjans, K.; Leitersdorf, E.; Fruchart, J.C. Mechanism of action of fibrates on lipid and lipoprotein metabolism. *Circulation* **1998**, *98*, 2088–2093. [CrossRef] [PubMed]

51. Frick, M.H.; Elo, O.; Haapa, K.; Heinonen, O.P.; Heinsalmi, P.; Helo, P.; Huttunen, J.K.; Kaitaniemi, P.; Koskinen, P.; Manninen, V.; et al. Helsinki heart study: Primary-prevention trial with gemfibrozil in middle-aged men with dyslipidemia. Safety of treatment, changes in risk factors, and incidence of coronary heart disease. *N. Engl. J. Med.* **1987**, *317*, 1237–1245. [CrossRef] [PubMed]

52. Rubins, H.B.; Robins, S.J.; Collins, D.; Fye, C.L.; Anderson, J.W.; Elam, M.B.; Faas, F.H.; Linares, E.; Schaefer, E.J.; Schectman, G.; et al. Gemfibrozil for the secondary prevention of coronary heart disease in men with low levels of high-density lipoprotein cholesterol. Veterans affairs high-density lipoprotein cholesterol intervention trial study group. *N. Engl. J. Med.* **1999**, *341*, 410–418. [CrossRef] [PubMed]

53. Elam, M.; Lovato, L.C.; Ginsberg, H. Role of fibrates in cardiovascular disease prevention, the accord-lipid perspective. *Curr. Opin. Lipidol.* **2011**, *22*, 55–61. [CrossRef] [PubMed]

54. Choi, H.D.; Shin, W.G. Safety and efficacy of statin treatment alone and in combination with fibrates in patients with dyslipidemia: A meta-analysis. *Curr. Med. Res. Opin.* **2014**, *30*, 1–10. [CrossRef] [PubMed]

55. Bloomfield, H.E. The role of fibrates in a statin world. *Arch. Intern. Med.* **2006**, *166*, 715–716. [CrossRef] [PubMed]

56. Keene, D.; Price, C.; Shun-Shin, M.J.; Francis, D.P. Effect on cardiovascular risk of high density lipoprotein targeted drug treatments niacin, fibrates, and CETP inhibitors: Meta-analysis of randomised controlled trials including 117,411 patients. *BMJ* **2014**, *349*, g4379. [CrossRef] [PubMed]

57. Fazio, S.; Linton, M.F. Debate: "How low should LDL cholesterol be lowered?" Viewpoint: "It doesn't need to be very low". *Curr. Control. Trials Cardiovasc. Med.* **2001**, *2*, 8–11. [CrossRef] [PubMed]

58. Abifadel, M.; Varret, M.; Rabes, J.P.; Allard, D.; Ouguerram, K.; Devillers, M.; Cruaud, C.; Benjannet, S.; Wickham, L.; Erlich, D.; et al. Mutations in PCSK9 cause autosomal dominant hypercholesterolemia. *Nat. Genet.* **2003**, *34*, 154–156. [CrossRef] [PubMed]

59. Sabatine, M.S.; Giugliano, R.P.; Wiviott, S.D.; Raal, F.J.; Blom, D.J.; Robinson, J.; Ballantyne, C.M.; Somaratne, R.; Legg, J.; Wasserman, S.M.; et al. Efficacy and safety of evolocumab in reducing lipids and cardiovascular events. *N. Engl. J. Med.* **2015**, *372*, 1500–1509. [CrossRef] [PubMed]

60. Chaudhary, R.; Garg, J.; Shah, N.; Sumner, A. PCSK9 inhibitors: A new era of lipid lowering therapy. *World J. Cardiol.* **2017**, *9*, 76–91. [CrossRef] [PubMed]

61. Lipinski, M.J.; Benedetto, U.; Escarcega, R.O.; Biondi-Zoccai, G.; Lhermusier, T.; Baker, N.C.; Torguson, R.; Brewer, H.B., Jr.; Waksman, R. The impact of proprotein convertase subtilisin-kexin type 9 serine protease inhibitors on lipid levels and outcomes in patients with primary hypercholesterolaemia: A network meta-analysis. *Eur. Heart J.* **2016**, *37*, 536–545. [CrossRef] [PubMed]

62. Raal, F.J.; Stein, E.A.; Dufour, R.; Turner, T.; Civeira, F.; Burgess, L.; Langslet, G.; Scott, R.; Olsson, A.G.; Sullivan, D.; et al. PCSK9 inhibition with evolocumab (AMG 145) in heterozygous familial hypercholesterolaemia (RUTHERFORD-2): A randomised, double-blind, placebo-controlled trial. *Lancet* **2015**, *385*, 331–340. [CrossRef]

63. Raal, F.J.; Honarpour, N.; Blom, D.J.; Hovingh, G.K.; Xu, F.; Scott, R.; Wasserman, S.M.; Stein, E.A.; Investigators, T. Inhibition of PCSK9 with evolocumab in homozygous familial hypercholesterolaemia (tesla Part B): A randomised, double-blind, placebo-controlled trial. *Lancet* **2015**, *385*, 341–350. [CrossRef]

64. Nicholls, S.J.; Puri, R.; Anderson, T.; Ballantyne, C.M.; Cho, L.; Kastelein, J.J.; Koenig, W.; Somaratne, R.; Kassahun, H.; Yang, J.; et al. Effect of evolocumab on progression of coronary disease in statin-treated patients: The glagov randomized clinical trial. *JAMA* **2016**, *316*, 2373–2384. [CrossRef] [PubMed]

65. Sabatine, M.S.; Giugliano, R.P.; Keech, A.C.; Honarpour, N.; Wiviott, S.D.; Murphy, S.A.; Kuder, J.F.; Wang, H.; Liu, T.; Wasserman, S.M.; et al. Evolocumab and clinical outcomes in patients with cardiovascular disease. *N. Engl. J. Med.* **2017**, *376*, 1713–1722. [CrossRef] [PubMed]

66. Toth, P.P.; Worthy, G.; Gandra, S.R.; Sattar, N.; Bray, S.; Cheng, L.I.; Bridges, I.; Worth, G.M.; Dent, R.; Forbes, C.A.; et al. Systematic review and network meta-analysis on the efficacy of evolocumab and other therapies for the management of lipid levels in hyperlipidemia. *J. Am. Heart Assoc.* **2017**, *6*. [CrossRef] [PubMed]

67. Giugliano, R.P.; Pedersen, T.R.; Park, J.G.; De Ferrari, G.M.; Gaciong, Z.A.; Ceska, R.; Toth, K.; Gouni-Berthold, I.; Lopez-Miranda, J.; Schiele, F.; et al. Clinical efficacy and safety of achieving very low LDL-cholesterol concentrations with the PCSK9 inhibitor evolocumab: A prespecified secondary analysis of the fourier trial. *Lancet* **2017**, *390*, 1962–1971. [CrossRef]

68. Giugliano, R.P.; Mach, F.; Zavitz, K.; Kurtz, C.; Im, K.; Kanevsky, E.; Schneider, J.; Wang, H.; Keech, A.; Pedersen, T.R.; et al. Cognitive function in a randomized trial of evolocumab. *N. Engl. J. Med.* **2017**, *377*, 633–643. [CrossRef] [PubMed]

69. Sabatine, M.S.; Leiter, L.A.; Wiviott, S.D.; Giugliano, R.P.; Deedwania, P.; De Ferrari, G.M.; Murphy, S.A.; Kuder, J.F.; Gouni-Berthold, I.; Lewis, B.S.; et al. Cardiovascular safety and efficacy of the PCSK9 inhibitor evolocumab in patients with and without diabetes and the effect of evolocumab on glycaemia and risk of new-onset diabetes: A prespecified analysis of the fourier randomised controlled trial. *Lancet Diabetes Endocrinol.* **2017**, *5*, 941–950. [CrossRef]

70. Bonaca, M.P.; Nault, P.; Giugliano, R.P.; Keech, A.C.; Pineda, A.L.; Kanevsky, E.; Kuder, J.; Murphy, S.A.; Jukema, J.W.; Lewis, B.S.; et al. Low-density lipoprotein cholesterol lowering with evolocumab and outcomes in patients with peripheral artery disease: Insights from the fourier trial (further cardiovascular outcomes research with PCSK9 inhibition in subjects with elevated risk). *Circulation* **2018**, *137*, 338–350. [CrossRef] [PubMed]

71. Ridker, P.M.; Tardif, J.C.; Amarenco, P.; Duggan, W.; Glynn, R.J.; Jukema, J.W.; Kastelein, J.J.P.; Kim, A.M.; Koenig, W.; Nissen, S.; et al. Lipid-reduction variability and antidrug-antibody formation with bococizumab. *N. Engl. J. Med.* **2017**, *376*, 1517–1526. [CrossRef] [PubMed]

72. Ridker, P.M.; Revkin, J.; Amarenco, P.; Brunell, R.; Curto, M.; Civeira, F.; Flather, M.; Glynn, R.J.; Gregoire, J.; Jukema, J.W.; et al. Cardiovascular efficacy and safety of bococizumab in high-risk patients. *N. Engl. J. Med.* **2017**, *376*, 1527–1539. [CrossRef] [PubMed]

73. Bandyopadhyay, D.; Hajra, A.; Ashish, K.; Qureshi, A.; Ball, S. New hope for hyperlipidemia management: Inclisiran. *J. Cardiol.* **2017**. [CrossRef] [PubMed]

74. Ray, K.K.; Landmesser, U.; Leiter, L.A.; Kallend, D.; Dufour, R.; Karakas, M.; Hall, T.; Troquay, R.P.; Turner, T.; Visseren, F.L.; et al. Inclisiran in patients at high cardiovascular risk with elevated LDL cholesterol. *N. Engl. J. Med.* **2017**, *376*, 1430–1440. [CrossRef] [PubMed]

75. Landmesser, U.; Chapman, M.J.; Stock, J.K.; Amarenco, P.; Belch, J.J.F.; Boren, J.; Farnier, M.; Ference, B.A.; Gielen, S.; Graham, I.; et al. 2017 update of ESC/EAS task force on practical clinical guidance for proprotein convertase subtilisin/kexin type 9 inhibition in patients with atherosclerotic cardiovascular disease or in familial hypercholesterolaemia. *Eur. Heart J.* **2017**. [CrossRef] [PubMed]

76. Norata, G.D.; Tavori, H.; Pirillo, A.; Fazio, S.; Catapano, A.L. Biology of proprotein convertase subtilisin kexin 9: Beyond low-density lipoprotein cholesterol lowering. *Cardiovasc. Res.* **2016**, *112*, 429–442. [CrossRef] [PubMed]

77. Ferri, N.; Tibolla, G.; Pirillo, A.; Cipollone, F.; Mezzetti, A.; Pacia, S.; Corsini, A.; Catapano, A.L. Proprotein convertase subtilisin kexin type 9 (PCSK9) secreted by cultured smooth muscle cells reduces macrophages ldlr levels. *Atherosclerosis* **2012**, *220*, 381–386. [CrossRef] [PubMed]

78. Wu, C.Y.; Tang, Z.H.; Jiang, L.; Li, X.F.; Jiang, Z.S.; Liu, L.S. PCSK9 sirna inhibits HUVEC apoptosis induced by ox-LDL via Bcl/Bax-caspase9-caspase3 pathway. *Mol. Cell. Biochem.* **2012**, *359*, 347–358. [CrossRef] [PubMed]

79. Sattar, N.; Preiss, D.; Murray, H.M.; Welsh, P.; Buckley, B.M.; de Craen, A.J.; Seshasai, S.R.; McMurray, J.J.; Freeman, D.J.; Jukema, J.W.; et al. Statins and risk of incident diabetes: A collaborative meta-analysis of randomised statin trials. *Lancet* **2010**, *375*, 735–742. [CrossRef]

80. Swerdlow, D.I.; Preiss, D.; Kuchenbaecker, K.B.; Holmes, M.V.; Engmann, J.E.; Shah, T.; Sofat, R.; Stender, S.; Johnson, P.C.; Scott, R.A.; et al. HMG-coenzyme a reductase inhibition, type 2 diabetes, and bodyweight: Evidence from genetic analysis and randomised trials. *Lancet* **2015**, *385*, 351–361. [CrossRef]

81. Strandberg, T.E.; Pienimaki, T.; Strandberg, A.Y.; Pitkala, K.H.; Tilvis, R.S. Association between use of statin medication and weight change in older men. *J. Am. Geriatr. Soc.* **2012**, *60*, 1588–1590. [CrossRef] [PubMed]

82. Ong, K.L.; Waters, D.D.; Messig, M.; DeMicco, D.A.; Rye, K.A.; Barter, P.J. Effect of change in body weight on incident diabetes mellitus in patients with stable coronary artery disease treated with atorvastatin (from the treating to new targets study). *Am. J. Cardiol.* **2014**, *113*, 1593–1598. [CrossRef] [PubMed]

83. Besseling, J.; Kastelein, J.J.; Defesche, J.C.; Hutten, B.A.; Hovingh, G.K. Association between familial hypercholesterolemia and prevalence of type 2 diabetes mellitus. *JAMA* **2015**, *313*, 1029–1036. [CrossRef] [PubMed]

84. Filippatos, T.D.; Filippas-Ntekouan, S.; Pappa, E.; Panagiotopoulou, T.; Tsimihodimos, V.; Elisaf, M.S. PCSK9 and carbohydrate metabolism: A double-edged sword. *World J. Diabetes* **2017**, *8*, 311–316. [CrossRef] [PubMed]

85. Cohen, J.C.; Boerwinkle, E.; Mosley, T.H., Jr.; Hobbs, H.H. Sequence variations in PCSK9, low LDL, and protection against coronary heart disease. *N. Engl. J. Med.* **2006**, *354*, 1264–1272. [CrossRef] [PubMed]

86. Awan, Z.; Delvin, E.E.; Levy, E.; Genest, J.; Davignon, J.; Seidah, N.G.; Baass, A. Regional distribution and metabolic effect of PCSK9 insleu and R46L gene mutations and apoe genotype. *Can. J. Cardiol.* **2013**, *29*, 927–933. [CrossRef] [PubMed]

87. Cao, Y.X.; Liu, H.H.; Dong, Q.T.; Li, S.; Li, J.J. Effect of proprotein convertase subtilisin/kexin type 9 (PCSK9) monoclonal antibodies on new-onset diabetes mellitus and glucose metabolism: A systematic review and meta-analysis. *Diabetes Obes. Metab.* **2018**. [CrossRef] [PubMed]

88. Verbeek, R.; Boyer, M.; Boekholdt, S.M.; Hovingh, G.K.; Kastelein, J.J.; Wareham, N.; Khaw, K.T.; Arsenault, B.J. Carriers of the PCSK9 R46L variant are characterized by an antiatherogenic lipoprotein profile assessed by nuclear magnetic resonance spectroscopy-brief report. *Arterioscler. Thromb. Vasc. Biol.* **2017**, *37*, 43–48. [CrossRef] [PubMed]

89. Kwakernaak, A.J.; Lambert, G.; Dullaart, R.P. Plasma proprotein convertase subtilisin-kexin type 9 is predominantly related to intermediate density lipoproteins. *Clin. Biochem.* **2014**, *47*, 679–682. [CrossRef] [PubMed]

90. Baragetti, A.; Grejtakova, D.; Casula, M.; Olmastroni, E.; Jotti, G.S.; Norata, G.D.; Catapano, A.L.; Bellosta, S. Proprotein convertase subtilisin-kexin type-9 (PCSK9) and triglyceride-rich lipoprotein metabolism: Facts and gaps. *Pharmacol. Res.* **2018**, *130*, 1–11. [CrossRef] [PubMed]

91. Ooi, T.C.; Krysa, J.A.; Chaker, S.; Abujrad, H.; Mayne, J.; Henry, K.; Cousins, M.; Raymond, A.; Favreau, C.; Taljaard, M.; et al. The effect of PCSK9 loss-of-function variants on the postprandial lipid and apob-lipoprotein response. *J. Clin. Endocrinol. Metab.* **2017**, *102*, 3452–3460. [CrossRef] [PubMed]

92. Reyes-Soffer, G.; Pavlyha, M.; Ngai, C.; Thomas, T.; Holleran, S.; Ramakrishnan, R.; Karmally, W.; Nandakumar, R.; Fontanez, N.; Obunike, J.; et al. Effects of PCSK9 inhibition with alirocumab on lipoprotein metabolism in healthy humans. *Circulation* **2017**, *135*, 352–362. [CrossRef] [PubMed]

93. Sniderman, A.D.; De Graaf, J.; Couture, P.; Williams, K.; Kiss, R.S.; Watts, G.F. Regulation of plasma LDL: The apob paradigm. *Clin. Sci.* **2009**, *118*, 333–339. [CrossRef] [PubMed]

94. Filippatos, T.D.; Kei, A.; Rizos, C.V.; Elisaf, M.S. Effects of PCSK9 inhibitors on other than low-density lipoprotein cholesterol lipid variables. *J. Cardiovasc. Pharmacol. Ther.* **2018**, *23*, 3–12. [CrossRef] [PubMed]

95. Koren, M.J.; Giugliano, R.P.; Raal, F.J.; Sullivan, D.; Bolognese, M.; Langslet, G.; Civeira, F.; Somaratne, R.; Nelson, P.; Liu, T.; et al. Efficacy and safety of longer-term administration of evolocumab (AMG 145) in patients with hypercholesterolemia: 52-week results from the open-label study of long-term evaluation against LDL-C (OSLER) randomized trial. *Circulation* **2014**, *129*, 234–243. [CrossRef] [PubMed]

96. Robinson, J.G.; Farnier, M.; Krempf, M.; Bergeron, J.; Luc, G.; Averna, M.; Stroes, E.S.; Langslet, G.; Raal, F.J.; El Shahawy, M.; et al. Efficacy and safety of alirocumab in reducing lipids and cardiovascular events. *N. Engl. J. Med.* **2015**, *372*, 1489–1499. [CrossRef] [PubMed]

97. Choi, S.; Korstanje, R. Proprotein convertases in high-density lipoprotein metabolism. *Biomark. Res.* **2013**, *1*, 27. [CrossRef] [PubMed]

98. Ferri, N.; Corsini, A.; Macchi, C.; Magni, P.; Ruscica, M. Proprotein convertase subtilisin kexin type 9 and high-density lipoprotein metabolism: Experimental animal models and clinical evidence. *Transl. Res.* **2016**, *173*, 19–29. [CrossRef] [PubMed]

99. Girona, J.; Ibarretxe, D.; Plana, N.; Guaita-Esteruelas, S.; Amigo, N.; Heras, M.; Masana, L. Circulating PCSK9 levels and CETP plasma activity are independently associated in patients with metabolic diseases. *Cardiovasc. Diabetol.* **2016**, *15*, 107. [CrossRef] [PubMed]

100. Hegele, R.A. PCSK9 inhibitors: Smooth sailing or a little turbulence ahead? *Lancet Diabetes Endocrinol.* **2017**, *5*, 490–492. [CrossRef]

101. Ference, B.A.; Robinson, J.G.; Brook, R.D.; Catapano, A.L.; Chapman, M.J.; Neff, D.R.; Voros, S.; Giugliano, R.P.; Davey Smith, G.; Fazio, S.; et al. Variation in PCSK9 and hmgcr and risk of cardiovascular disease and diabetes. *N. Engl. J. Med.* **2016**, *375*, 2144–2153. [CrossRef] [PubMed]

102. Jellinger, P.S.; Handelsman, Y.; Rosenblit, P.D.; Bloomgarden, Z.T.; Fonseca, V.A.; Garber, A.J.; Grunberger, G.; Guerin, C.K.; Bell, D.S.H.; Mechanick, J.I.; et al. American association of clinical endocrinologists and american college of endocrinology guidelines for management of dyslipidemia and prevention of cardiovascular disease. *Endocr. Pract.* **2017**, *23*, 1–87. [CrossRef] [PubMed]

103. Catapano, A.L.; Graham, I.; De Backer, G.; Wiklund, O.; Chapman, M.J.; Drexel, H.; Hoes, A.W.; Jennings, C.S.; Landmesser, U.; Pedersen, T.R.; et al. 2016 ESC/EAS guidelines for the management of dyslipidaemias. *Eur. Heart J.* **2016**, *37*, 2999–3058. [CrossRef] [PubMed]

104. Olsson, A.G.; Angelin, B.; Assmann, G.; Binder, C.J.; Bjorkhem, I.; Cedazo-Minguez, A.; Cohen, J.; von Eckardstein, A.; Farinaro, E.; Muller-Wieland, D.; et al. Can LDL cholesterol be too low? Possible risks of extremely low levels. *J. Intern. Med.* **2017**, *281*, 534–553. [CrossRef] [PubMed]

105. Radisauskas, R.; Kuzmickiene, I.; Milinaviciene, E.; Everatt, R. Hypertension, serum lipids and cancer risk: A review of epidemiological evidence. *Medicina* **2016**, *52*, 89–98. [CrossRef] [PubMed]
106. Benn, M.; Tybjaerg-Hansen, A.; Stender, S.; Frikke-Schmidt, R.; Nordestgaard, B.G. Low-density lipoprotein cholesterol and the risk of cancer: A mendelian randomization study. *J. Natl. Cancer Inst.* **2011**, *103*, 508–519. [CrossRef] [PubMed]

Perspective

Epigenetic Regulation of ATP-Binding Cassette Protein A1 (*ABCA1*) Gene Expression: A New Era to Alleviate Atherosclerotic Cardiovascular Disease

Mohamed Zaiou [1,*] **and Ahmed Bakillah** [2]

[1] School of Pharmacy, University of Lorraine, 5 rue Albert Lebrun, 54000 Nancy, France
[2] Department of Medicine, Downstate Medical Center, State University of New York, 450 Clarkson Ave, Brooklyn, NY 11203, USA; Ahmed.Bakillah@downstate.edu
* Correspondence: mohamed.zaiou@univ-lorraine.fr; Tel.: +33-0-372-747-300; Fax: +33-0-383-682-301

Received: 16 April 2018; Accepted: 2 May 2018; Published: 3 May 2018

Abstract: The most important function of high density lipoprotein (HDL) is its ability to remove cholesterol from cells and tissues involved in the early stages of atherosclerosis back to the liver for excretion. The ATP-binding cassette transporters ABCA1 and ABCG1 are responsible for the major part of cholesterol efflux to HDL in macrophage foam cells. Thus, promoting the process of reverse cholesterol transport (RCT) by upregulating mainly ABCA1 remains one of the potential targets for the development of new therapeutic agents against atherosclerosis. Growing evidence suggests that posttranscriptional regulation of HDL biogenesis as well as modulation of ABCA1 expression are under the control of several genetic and epigenetic factors such as transcription factor (TFs), microRNAs (miRNAs) and RNA-binding proteins (RBPs).These factors may act either individually or in combination to orchestrate ABCA1 expression. Complementary to our recent work, we propose an exploratory model for the potential molecular mechanism(s) underlying epigenetic signature of ABCA1 gene regulation. Such a model may hopefully provide the basic framework for understanding the epigenetic regulation of RCT and contribute to the development of novel therapeutic strategies to alleviate the burden of cardiovascular diseases (CVD).

Keywords: ABCA1; HDL; miRNA; circular RNA; gene expression; RNA-binding proteins; reverse cholesterol transport; cardiovascular diseases

---

1. Introduction

Based on the general consensus that HDL protects against atherosclerotic CVD, several attempts have been made to design drugs that raise HDL cholesterol (HDL-C) levels or enhance its cardioprotective function. However, several prospective HDL-C-raising trials have failed to demonstrate improved efficacy for major adverse cardiovascular outcomes. In this prospect, there are continuous efforts to increase HDL functionality to enhance RCT and potentially achieve further cardiovascular event reduction.

Over the last decades, genetic breakthroughs have revolutionized cardiovascular research by providing opportunities to elucidate novel molecular mechanisms underlying HDL biogenesis and RCT pathways. Some preclinical and clinical studies have shown that apoA-I Milano, a naturally occurring mutant of ApoA-I (Arg173Cysteine) has beneficial athero-protective and anti-inflammatory effects [1]. Furthermore, research has suggested that the expression of genes involved in the RCT combines several complex regulatory networks that are controlled by at least two types of trans-factors: transcription factors (TFs) in the nucleus and posttranscriptional epigenetic factors that bind to cis-regulatory RNA elements mostly located in the 3′UTR of their target mRNAs. Epigenetic factors including miRNAs and RBPs have been shown to be associated with several pathophysiological conditions. For instance, the dysregulation of miRNAs has been associated with the disruption of multiple gene networks leading to metabolic disorders including diabetes, obesity, metabolic syndrome and atherosclerosis [2,3]. This clearly indicates that beyond the existing therapies, advance in epigenetics mechanisms could offer additional opportunities to develop novel treatment strategies for atherosclerosis. In this regard, attempts have been made to develop Apabetalone (RVX-208) as the first epigenetic approach to treat CVD. In recent clinical trials such as SUSTAIN and ASSURE, RVX-208, an orally active molecule has been shown to increase plasma apoA-I, HDL-C (pre-beta and alpha-HDL), and enhance the ability of serum to efflux cholesterol via ABCA1, ABCG1, and scavenger receptor class B type I (SR-BI)-dependent pathways [4]. RVX-208 increases apoA-I transcription through an epigenetic mechanism by inhibiting the bromodomain and extra-terminal domain (BET) protein 4 (BRD4) [5]. Most importantly, emerging evidence has suggested that miRNAs act as a novel class of epigenetic regulators of RCT and HDL-C from synthesis to clearance, and thus contributing remarkably to the pathogenesis of atherosclerosis [6]. Hence, understanding epigenetic mechanisms that control RCT genes network, mainly *ABCA1* gene that is involved in the initiation of this process, may bring insights into novel therapeutic approaches for treating atherosclerotic vascular disease.

2. Discussion

ABCA1 is a key mediator of cholesterol efflux to lipid-poor apolipoprotein A-1. Several studies have confirmed that compromised ABCA1 activity leads to accelerated and early atherogenesis. As summarized in our recent paper [6], the RCT gene network is a complex process highly controlled at both transcriptional and posttranscriptional levels. Herein, we consider the *ABCA1* gene, for which some experimental knowledge is available both at the transcriptional and posttranscriptional levels, as an example to illustrate potential epigenetic mechanisms driving the regulation of RCT. We propose a hypothetical model for the potential dynamic interplay between different genetic and epigenetic regulators that may serve to regulate *ABCA1* gene enabling the cell to respond to different environment changes (Figure 1). However, all proposed axes of interaction, if they occur, need to be experimentally validated in different models.

At the transcriptional level, several nuclear receptors including peroxisome proliferator-activated receptors (PPARs), liver X-receptor (LXR), and farnesoid X receptor (FXR), have been shown to influence lipid metabolism along with genes involved in RCT pathway including *ABCA1*. Unfortunately, despite the effectiveness of these nuclear receptors in preclinical studies, their translation to human clinical trials is still facing many challenges.

At the posttranscriptional level, the 3′-UTR of *ABCA1* gene has been shown to be directly targeted by multiple miRNAs including miR-33, miR-758, miR-145, miR-27, miR-144, miR-26 and miR-106, which lead to cholesterol efflux and HDL-C levels repression [6]. In addition to miRNAs, RNA-binding proteins (RBPs) are also known to bind to AU-rich elements (AREs) in the 3′UTR of many genes, and thereby modulate their expression by increasing or decreasing mRNAs' translation and/or stability. With respect to ABCA1, human antigen R (HuR), an RBP, has been reported to bind to the 3′-UTR of this transporter and increase its expression by enhancing protein translation [7]. Based on this information, it would be tempting to suggest the existence of a possible regulatory relationship between miRNAs and RBPs that could influence ABCA1 expression. In this context, specific miRNAs and RBPs may act via cooperation/competition to directly or indirectly regulate gene expression. Furthermore, post-translational modifications of RBPs, including their phosphorylation and methylation, provide additional layers of complexity, as they control RNA-binding, function and localization [8]. Therefore, phosphorylation and nuclear transit of RBPs could be another possible mechanism to influence RBP-mediated regulation of *ABCA1* gene expression as suggested in a previous study [9]. In addition to these potential mechanisms, certain miRNA species may also control the expression of other important epigenetic regulators such as DNA methyltransferases and histone deacetylases. Conversely, DNA/RNA methylation and histone modification may contribute to the regulation of these miRNAs

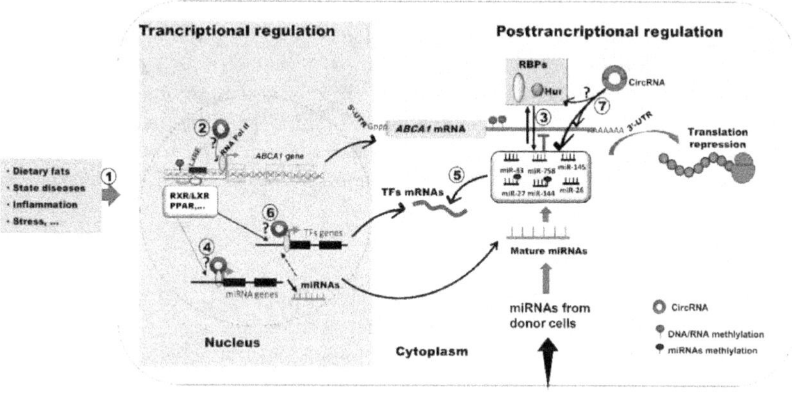

Abbreviations: TFs, transcription factors, RBPs, RNA-binding proteins, Hur: Human antigen R. CircRNA, circular RNA, ABCA1, ATP-binding cassette transporter A1, RXR, Retinoid X receptor, LXR, Liver X receptor, PPAR, peroxisome proliferator-activated receptor.

Figure 1. Schematic representation of proposed scenarios for the mechanisms by which *ABCA1* gene expression is regulated. Various factors including nutrition, state diseases, stress and inflammation can alter epigenetic *ABCA1* gene regulation in different cells mainly macrophages and hepatocytes (**1**). At the transcriptional level, *ABCA1* gene expression is regulated by key nuclear receptors including LXR family and their heterodimeric partners, retinoic acid receptors (RXR) via functional LXREs (**2**). At the post-transcriptional level, miRNAs and RBPs (Hur) cooperate/compete for target binding regulation (**3**). miRNAs and TFs may cooperate to tune gene expression by forming feedback or feedforward loops (**4**) and (**5**). CircRNAs may also be part of the interplay by their potential interaction with TFs, miRNAs and RBPs (**2**), (**4**), (**6**) and (**7**). Finally, the proposed interplay between TFs-miRNAs-RBPs-CircRNAs could be taken into consideration while elucidating epigenetic mechanisms regulating the first step of RCT gene network.

As is true for protein-coding genes, the expression of miRNAs is also under the control of numerous transcription factors. Aberrant regulation of miRNAs by TFs can cause phenotypic variations and diseases. As an example, miR-26, known to directly target the *ABCA1* 3′-UTR, thereby repressing cholesterol efflux and HDL-C levels, has been shown to be inhibited in cells treated with LXR agonists [10]. Moreover, the activation of LXR has been associated with an increase in miR-144 and ABCA1 expression to fine-tune RCT by macrophages [11]. On the other hand, miRNAs/TFs feedback-loop regulation may occur and represent another possible mechanism involved in the regulation of the RCT gene networks [12]. As an example of the TFs–miRNA regulatory network, RXRα has been reported to be regulated by miRNAs including miR-128-2 [13], while other miRNAs have been shown to regulate the expression and activity of different TFs such PPARs. Hence, it is reasonable to assume the existence of a possible cross-talk between TFs and miRNAs that could modulate *ABCA1* gene and RCT circuit. However, such mechanisms of regulation are still not well defined and could represent an important future area of research.

Circular RNAs (circRNAs), a class of RNAs with the linking of 3′ and 5′ ends, are predicted to function as robust transcriptional and posttranscriptional regulators of gene expression. Evidence is arising that some circRNAs might regulate miRNA function as microRNA sponges. Relevant information in this respect can be provided by the web tool CircInteractome, freely available at http://circinteractome.nia.nih.gov. This new database has been developed for exploring circRNAs and their interacting proteins and microRNAs [14]. Further studies have suggested that circRNA-miRNA-mRNA axes play a prominent role in different pathologies including CVD [15]. Some circRNAs have been predominantly localized in the nucleus where they can directly promote host-gene transcription through interaction with RNA polymerase II (pol II) in the promoter region of genes. Furthermore, these RNAs may interact with TFs to influence various diseases [16].

Because no data are yet available regarding the impact of circRNAs on lipid homeostasis and HDL biogenesis, this has prompted us to propose a model integrating TFs, miRNAs, RBPs and CirRNAs into the *ABCA1* regulatory gene network (Figure 1). However, we are aware that certain circuits of this model are much more difficult to directly translate in vivo mostly due to the complexity of several highly connected pathways with feedback and feedforward loops. Thus, to validate the proposed epigenetic transcriptional and post-transcriptional mechanisms for *ABCA1* gene expression regulation, future well-designed in vivo experimental studies combined with high-throughput sequencing and bioinformatic data are required.

3. Conclusions and Perspectives

Emerging research suggests that deregulated miRNAs can impact RCT gene networks, mainly *ABCA1* gene. These observations have generated a renewed interest in novel targets for epigenetic regulators that may pave the way for establishing novel strategies to raise functional HDL and promote RCT. However, since the field of epigenetic regulation by small RNAs is still in its infancy, there are only a few experimentally validated data associating the dysregulation of miRNAs with genes involved in the HDL biogenesis and RCT. Our proposed model (Figure 1) for potential epigenetic regulation of *ABCA1* is expected to be translated into specific questions and hypothesis which may benefit further research in this area. However, one of the greatest challenges one may confront here is how to fully gain knowledge from the potential interplay and/or cross-talk between various miRNAs and other cellular regulatory factors controlling specific pathway(s) of cholesterol removal from cells and how all these networks can shape a physiological function in both normal and pathological cells in response to a behavior or a stimulus. To answer these questions, more focus should be given to the integration of computational approaches such as systems biology and molecular networks modeling in combination with high-throughput experiments to provide a list of potential miRNA target genes involved in the RCT process. This might help us to better understand the consequences of the complex interplay of miRNAs with other epigenetic regulators on *ABCA1* gene and possibly other relevant genes involved in RCT regulation.

Author Contributions: M.Z. and A.B. conception, design of perspectives, wrote, read and approved the manuscript.

Conflicts of Interest: The authors declare no conflicts of interest.

References

1. Chyu, K.Y.; Shah, P.K. HDL/ApoA-1 infusion and ApoA-1 gene therapy in atherosclerosis. *Front. Pharmacol.* **2015**, *6*, 187. [CrossRef] [PubMed]
2. Rottiers, V.; Näär, A.M. MicroRNAs in metabolism and metabolic disorders. *Nat. Rev. Mol. Cell Biol.* **2012**, *13*, 239–250. [CrossRef] [PubMed]
3. Zaiou, M.; El Amri, H.; Bakillah, A. The clinical potential of adipogenesis and obesity-related microRNAs. *Nutr. Metab. Cardiovasc. Dis.* **2018**, *28*, 91–111. [CrossRef] [PubMed]
4. Gilham, D.; Wasiak, S.; Tsujikawa, L.M.; Halliday, C.; Norek, K.; Patel, R.G.; Kulikowski, E.; Johansson, J.; Sweeney, M.; Wong, N.C. RVX-208, a BET-inhibitor for treating atherosclerotic cardiovascular disease, raises ApoA-I/HDL and represses pathways that contribute to cardiovascular disease. *Atherosclerosis* **2016**, *247*, 48–57. [CrossRef] [PubMed]
5. Picaud, S.; Wells, C.; Felletar, I.; Brotherton, D.; Martin, S.; Savitsky, P.; Diez-Dacal, B.; Philpott, M.; Bountra, C.; Lingard, H.; et al. RVX-208, an inhibitor of BET transcriptional regulators with selectivity for the second bromodomain. *Proc. Natl. Acad. Sci. USA* **2013**, *110*, 19754–19759. [CrossRef] [PubMed]
6. Zaiou, M.; Rihn, B.; Bakillah, A. Epigenetic regulation of genes involved in the reverse cholesterol transport through interaction with miRNAs. *Front. Biosci. Landmark* **2018**, *23*, 2090–2105.
7. Ramírez, C.M.; Lin, C.S.; Abdelmohsen, K.; Goedeke, L.; Yoon, J.H.; Madrigal-Matute, J.; Martin-Ventura, J.L.; Vo, D.T.; Uren, P.J.; Penalva, L.O.; et al. RNA binding protein HuR regulates the expression of ABCA1. *J. Lipid Res.* **2014**, *55*, 1066–1076. [CrossRef] [PubMed]

8.   Glisovic, T.; Bachorik, J.L.; Yong, J.; Dreyfuss, G. RNA-binding proteins and post-transcriptional gene regulation. *FEBS Lett.* **2008**, *582*, 1977–1986. [CrossRef] [PubMed]

9.   Kurischko, C.; Broach, J.R. Phosphorylation and nuclear transit modulate the balance between normal function and terminal aggregation of the yeast RNA-binding protein Ssd1. *Mol. Biol. Cell.* **2017**, *28*, 3057–3069. [CrossRef] [PubMed]

10.   Sun, D.; Zhang, J.; Xie, J.; Wei, W.; Chen, M.; Zhao, X. MiR-26 controls LXR-dependent cholesterol efflux by targeting ABCA1 and ARL7. *FEBS Lett.* **2012**, *586*, 1472–1479. [CrossRef] [PubMed]

11.   Ramirez, C.M.; Rotllan, N.; Vlassov, A.V.; Dávalos, A.; Li, M.; Goedeke, L.; Aranda, J.F.; Cirera-Salinas, D.; Araldi, E.; Salerno, A.; et al. Control of cholesterol metabolism and plasma high-density lipoprotein levels by microRNA-144. *Circ. Res.* **2013**, *112*, 1592–1601. [CrossRef] [PubMed]

12.   Martinez, N.J.; Walhout, A.J. The interplay between transcription factors and microRNAs in genome-scale regulatory networks. *Bioessays* **2009**, *31*, 435–445. [CrossRef] [PubMed]

13.   Adlakha, Y.K.; Khanna, S.; Singh, R.; Singh, V.P.; Agrawal, A.; Saini, N. Pro-apoptotic miRNA-128-2 modulates ABCA1, ABCG1 and RXRα expression and cholesterol homeostasis. *Cell Death Dis.* **2013**, *4*, e780. [CrossRef] [PubMed]

14.   Dudekula, D.B.; Panda, A.C.; Grammatikakis, I.; De, S.; Abdelmohsen, K.; Gorospe, M. CircInteractome: A web tool for exploring circular RNAs and their interacting proteins and microRNAs. *RNA Biol.* **2016**, *13*, 34–42. [CrossRef] [PubMed]

15.   Holdt, L.M.; Stahringer, A.; Sass, K.; Pichler, G.; Kulak, N.A.; Wilfert, W.; Kohlmaier, A.; Herbst, A.; Northoff, B.H.; Nicolaou, A.; et al. Circular non-coding RNA ANRIL modulates ribosomal RNA maturation and atherosclerosis in humans. *Nat. Commun.* **2016**, *7*, 12429. [CrossRef] [PubMed]

16.   Yang, Z.G.; Awan, F.M.; Du, W.W.; Zeng, Y.; Lyu, J.; Wu, D.; Gupta, S.; Yang, W.; Yang, B.B. The Circular RNA Interacts with STAT3, Increasing Its Nuclear Translocation and Wound Repair by Modulating Dnmt3a and miR-17 Function. *Mol. Ther.* **2017**, *25*, 2062–2074. [CrossRef] [PubMed]

Review

Regulation of Sphingolipid Metabolism by MicroRNAs: A Potential Approach to Alleviate Atherosclerosis

Zainab Jahangir [1], **Ahmed Bakillah** [2] **and Jahangir Iqbal** [2,*]

[1]   John P. Stevens High School, North Edison, NJ 08820, USA; 3006676@edison.k12.nj.us
[2]   King Abdullah International Medical Research Center, Ministry of National Guard Health Affairs,
     Al Ahsa 31982, Saudi Arabia; bakillahah@ngha.med.sa
*   Correspondence: iqbalja@ngha.med.sa; Tel.: +966-56-433-6270; Fax: +966-13-533-9998

Received: 2 September 2018; Accepted: 17 September 2018; Published: 17 September 2018

Abstract: The rapidly expanding field of bioactive lipids is exemplified by the many sphingolipids, which are structurally and functionally diverse molecules with significant physiologic functions. These sphingolipids are main constituents of cellular membranes and have been found associated with plasma lipoproteins, and their concentrations are altered in several metabolic disorders such as atherosclerosis, obesity, and diabetes. Understanding the mechanisms that regulate their biosynthesis and secretion may provide novel information that might be amenable to therapeutic targeting in the treatment of these diseases. Several sphingolipid synthesis genes have been targeted as potential therapeutics for atherosclerosis. In recent years, significant progress has been made in studying the role of microRNAs (miRNAs) in lipid metabolism. However, little effort has been made to investigate their role in sphingolipid metabolism. Sphingolipid biosynthetic pathways involve various enzymes that lead to the formation of several key molecules implicated in atherosclerosis, and the identification of miRNAs that regulate these enzymes could help us to understand these complex pathways better and may prove beneficial in alleviating atherosclerosis.

Keywords: atherosclerosis; ceramides; lipids; lipoproteins; miRNA; sphingolipids; sphingomyelin

---

1. Introduction

High plasma lipid levels are major risk factors for several cardiovascular and metabolic disorders such as atherosclerosis, obesity, and diabetes. Some of the most important risk factors for atherosclerosis are the circulating levels of low-density lipoprotein (LDL) and high-density lipoprotein (HDL) cholesterol [1,2]. Besides traditional risk factors, changes in sphingolipids may contribute to the pathogenesis of cardiovascular disease (CVD) [3,4]. Sphingolipids are a class of lipids that contain a sphingoid base, an aliphatic amino alcohol including sphingosine. Sphingoid bases such as dihydrosphingosine and sphingosine are the fundamental building blocks of all sphingolipids. Sphingolipids are biologically active cell components that regulate cellular processes and play an important role in signal transduction and cellular stress responses. The synthesis and degradation of sphingolipids, which serve as both structural lipids as well as signaling molecules, are regulated to maintain homeostasis [3]. Sphingolipids are either derived from other sphingolipids through catabolism via the salvage pathway or synthesized de novo in the endoplasmic reticulum [5]. Ceramide is a simple sphingolipid composed of sphingosine and a fatty acid. Ceramide constitutes the hydrophobic backbone and serves as the key precursor for the de novo synthesis of other biologically active complex sphingolipids (Figure 1) [6]. It represents a nodal point in the sphingolipid de novo pathway. It can be glycosylated or acquire a polar head group to form glycosphingolipids or sphingomyelin (SM), respectively [3]. It can also be reversibly degraded to form sphingosine which in turn can be phosphorylated by sphingosine kinase (SPK) to form sphingosine-1-phospate (S1P). Once synthesized, sphingolipids can be transported to plasma lipoproteins [7] or remain associated with cellular membranes. An imbalance of sphingolipid levels in plasma and tissue is associated with several metabolic diseases including atherosclerosis [4].

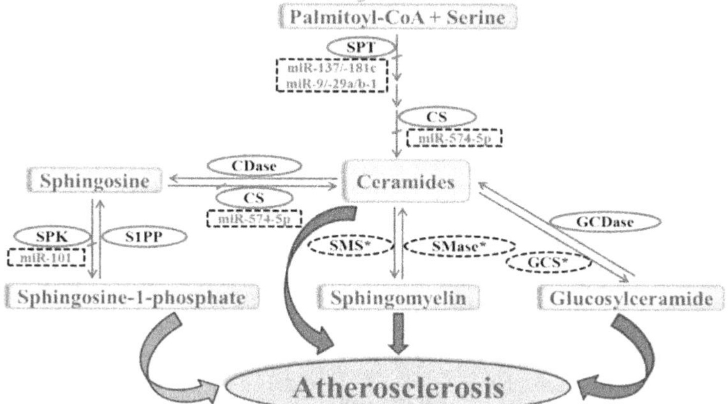

Figure 1. Schematic representation of miRNAs implicated in the synthesis of some key sphingolipid molecules. The figure is representative of some key enzymatic steps involved in sphingolipid biosynthetic pathways that are known to be regulated by various miRNAs (dotted rectangles). Enzymes with asterisks (dotted ovals) may be potential targets for other miRNAs that need to be identified. Increased levels of ceramides, sphingomyelin and glucosylceramide (red arrows) and decreased levels of sphingosine-1-phosphate (green arrow) have been implicated in atherosclerosis. The identification of novel miRNAs that regulate sphingolipid metabolism may be a potential therapeutic target to treat atherosclerosis. Abbreviations: SPT, serine palmitoyl transferase; CDase, ceramidase; CS, ceramide synthase; SPK, sphingosine kinase; S1PP, sphingosine-1-phosphate phosphatase; SMS, sphingomyelin synthase; SMase, sphingomyelinase; GCS, glucosylceramide synthase; GCDase, glucosylceramidase.

An increased plasma SM level has been proposed as an independent risk factor for coronary heart disease and has been shown to be associated with increased atherosclerosis in humans [8]. A reduction in SM levels in sphingomyelin synthase (SMS) knockout mice is known to decrease atherosclerosis [9,10]. Like SM, increased plasma and aortic ceramide levels are also associated with an increased risk of CVD [11]. Genetic deficiency or the inhibition of type 2-neutral sphingomyelinase (nSMase2), a key enzyme in sphingolipid metabolism, has been shown to decrease atherosclerosis in ApoE knockout mice by reducing inflammatory responses due to a decrease in ceramide levels [12]. Furthermore, plasma glycosphingolipid concentrations have been reported to be elevated in patients at increased risk of atherosclerosis [13]. The pharmacological inhibition of glucosylceramide synthase that is responsible for the synthesis of glucosylceramides has been reported to ameliorate atherosclerosis in ApoE knockout mice and rabbits [14]. In contrast to ceramide, glycosphingolipid, and SM, plasma S1P is believed to be cardioprotective [3]. Regulation of the interconvertible sphingolipid metabolites, ceramide and S1P, and their opposing signaling pathways may determine the net biological effect, a concept referred to as the "sphingolipid rheostat". Plasma S1P levels significantly decrease after myocardial infarction [15] and increase in patients after percutaneous coronary intervention [16]. It is believed that S1P bound to HDL may predict the severity of coronary heart disease [17]. Recently, it has been shown that apoM acts as a carrier and modulator of S1P that largely affects its homeostasis [18].

Lipid metabolism is a multi-faceted process that involves synthesis, accumulation, secretion, distribution to various tissues, degradation as well as excretion. Lipid metabolism is regulated by finely modulating a set of rate-limiting enzymes and transporters based on the needs of the cells [19]. A variety of cellular regulators including transcription factors that are involved in its synthesis and degradation are responsible for maintaining lipid homeostasis. The discovery that the cellular lipid and lipoprotein metabolism is targeted by miRNAs has provided new insight into the molecular mechanism of the pathogenesis of atherosclerosis, diabetes and obesity [20]. This has led to the exploitation of miRNAs as regulators of lipid metabolism with subsequent potential use in the treatment of lipid metabolism-related disorders [21,22]. This mini-review provides an overview of what is known about the regulation of lipid metabolism in general and sphingolipid metabolism in particular by miRNAs and emphasize their role in atherosclerosis.

2. Regulation of Lipid and Lipoprotein Metabolism by miRNAs

miRNAs are a class of small ~22-nucleotide long non-coding sequences which post-transcriptionally regulate gene expression to modulate a wide spectrum of biological processes [23,24]. They act as negative post-transcriptional regulators of gene expression [21,22] and have been proposed to regulate the expression of more than half of the human genes via controlling the expression of mRNA targets. miRNAs repress protein expression by mRNA destabilization and/or translational inhibition after binding the complementary sequences in the 3′ untranslated region (3′-UTR) of target mRNAs via seed sequences (Table 1) [25]. Individual miRNAs can target several mRNAs that are often connected in the same metabolic pathway [26]. Moreover, several miRNAs may regulate a single mRNA, thus enabling a fine-tuning of the targeted mRNA expression, to influence the regulation of cellular events [23,26]. Dysregulation of miRNAs has been correlated with disease pathogenesis [27]. Furthermore, miRNAs can be released from cells as exosomes in many body fluids and may serve as noninvasive biomarkers of diseases [28].

Table 1. Predicted seed sequence, target genes and tissues of miRNAs implicated in lipid and sphingolipid metabolism.

miRNAs	Predicted Seed Sequence *	Target Genes	Target Tissues
miR-33	GUUACGU	ABCA1, CROT, CPT1A, HADHB, ACLY, SREBF1, ACACA	Liver
miR-144	UAUGACA	ABCA1	Liver
miR-758	CAGUGUU	ABCA1	Liver
miR-26	AUGAACU	ABCA1	Liver
miR-106b	CGUGAAA	ABCA1	Liver
miR-27	AUUCGAG	SR-B1	Liver
miR-185	AGAGAGG	SR-B1	Liver
miR-96	CACGGUU	SR-B1, ABCA1	Liver
miR-223	UUGACUG	SR-B1	Liver
miR-30c	CAAAUG	MTTP, LPGAT1, ELOVL5, STARD3, MBOAT1	Liver
miR-128-1	GCCGGGG	LDLR	Liver
miR-148	ACGUGAC	LDLR, ABCA1	Liver
miR-122	UGUGAGG	FASN, SCD1, ACLY, ACC2	Liver
miR-155	CGUAAU	LXRα	Liver
miR-574	GUGUGAG	CerS	Multiple human cancer cells
miR-9	UGGUUUC	SPTLC1, SPTLC2	Primary astrocytes
miR-29a	UUUAGUC	SPTLC1, SPTLC2	Primary astrocytes
miR-29b-1	UUUGGUC	SPTLC1, SPTLC2	Primary astrocytes
miR-101	CUAUUGA	SPK	Colorectal cancer cells

* Conserved seed sequence for each miRNA was predicted by using TargetScan Human database (http://www.targetscan.org/vert_72/).

Over the past few decades, the role of miRNAs in controlling the cellular lipid and lipoprotein metabolism has been extensively studied. A large array of miRNAs participate in the lipid and lipoprotein metabolism by targeting enzymes and proteins that are involved in these metabolic pathways. This discovery has provided new insight into the biology and pathophysiology of CVD such as atherosclerosis [20,29]. Higher levels of HDL are inversely correlated with developing CVD. A number of miRNAs, of which miR-33a and miR-33b are the most well-studied, regulate HDL metabolism and control circulating cholesterol levels by regulating HDL synthesis, efflux and clearance [30,31]. miR-33a and miR-33b target multiple genes that regulate cholesterol efflux, which is regarded as an important cholesterol regulatory mechanism [32]. Studies in various animal models and cell lines have confirmed that miR-33 targets ATP-binding cassette (ABC) transporter A1, *ABCA1*, by binding to *3′-UTR* of *ABCA1* via seed sequence to regulate cholesterol metabolism by attenuating circulating HDL levels and thereby decreasing cholesterol efflux to apolipoprotein A1 (apoA1) [30,32]. Conversely, the silencing of miR-33 by using lentiviral expression clone containing anti-sense to miR-33 in experimental animal models generally results in increased hepatic expression of *ABCA1* and enhanced cholesterol efflux to ApoA1, thus increasing circulating HDL levels [32].

In vivo delivery of other miRNAs such as miR-144, miR-758, miR-26, and miR-106b by using either adenoviral or lentiviral expression clones has been shown to have similar results of decreased ABCA1 expression and reduced levels of circulating HDL-cholesterol [33]. The post-transcriptional repression of scavenger receptor class B type 1 (SR-B1) by miR-27, miR-185, miR-96, and miR-223 has also been shown to reduce selective HDL-cholesterol uptake [33]. Reverse cholesterol transport decreases cholesterol levels in peripheral macrophages and in atherosclerotic plaques to increase plaque stability and inhibit atherosclerosis progression [34]. These studies show that HDL metabolism provides a potential therapeutic target to treat atherosclerosis by regulating *ABCA1* via miRNAs.

Besides the HDL metabolism, reports have shown that miRNAs such as miR-30c, miR128-1, or miR-148a are also involved in controlling plasma LDL-cholesterol levels by regulating genes involved in cholesterol biosynthesis, very low-density lipoprotein (VLDL) secretion, and hepatic LDL receptor expression [30]. Recently, it was shown that miR-30c targets microsomal triglyceride transfer protein (MTTP) to regulate VLDL biogenesis and modulate lipid substrate availability for VLDL assembly by targeting genes involved in lipid biosynthesis such as lysophosphatidylglycerol acyltransferase 1 (LPGAT1), ELOVL fatty acid elongase 5 (ELOVL5), stAR related lipid transfer domain containing 3 (STARD3), and membrane bound *O*-acyltransferase domain containing 1 (MBOAT1) [35]. Inhibition of miR-122 has been shown to increase fatty acid oxidation and inhibit lipid synthesis genes such as fatty acid synthase (FASN), steroyl-coA desaturase 1 (SCD1), ATP citrate lyase (ACLY) and acetyl-coA carboxylase 2 (ACC2) and thereby reduce the availability of lipid substrates for VLDL biogenesis [36]. Similarly, inhibition of miR-33a and miR-33b expression in the liver is shown to increase fatty acid oxidation by increasing the expression of carnitine *O*-octanoyltransferase (CROT), carnitine palmitoyltransferase 1A (CPT1A) and hydroxyacyl-CoA dehydrogenase trifunctional multienzyme complex subunit beta (HADHB), and decrease its synthesis by reducing the synthesis of FASN, ACLY, sterol regulatory element binding factor 1 (SREBF1) and acetyl-CoA carboxylase alpha (ACACA) to reduce VLDL secretion in the circulation [37]. Furthermore, reports have suggested that increased expression of miR-155 in liver macrophages regulates fatty acid metabolism by directly targeting liver X receptor alpha (LXRα) and thereby reducing hepatic lipid accumulation [38]. These studies indicate that miRNAs exert their regulatory impacts at different levels of lipoprotein biosynthesis.

3. Regulation of Sphingolipid Metabolism by miRNAs

Sphingolipids have been recognized to regulate distinct biological functions beyond their role as structural membrane components. Within the past few decades, significant progress has been made toward understanding the role of sphingolipid pathways for atherosclerosis. The role of miRNAs to regulate sphingolipid metabolism has not been widely studied. There are only a few limited studies, done mostly in cancer cell lines, that have investigated the role of miRNAs in the regulation of enzymes involved in sphingolipid biosynthetic pathways (Figure 1) [39–41]. Ceramide is a central molecule in sphingolipid metabolism [6] which is generated either through de novo pathway or salvage pathway [5]. Ceramide acts as a signaling molecule, regulating many cellular responses and functions that may be involved in molecular mechanisms of CVD. The generation of ceramide can be significantly enhanced in certain inflammatory conditions such as atherosclerosis [12]. Ceramide is synthesized by a family of six ceramide synthases (CerS), each of which synthesizes ceramide with distinct acyl chain lengths. An alternatively spliced variant of ceramide synthase, CerS1-2, which is responsible for synthesizing C18-ceramide, is a target for miR-574-5p [41]. The knockdown of miR-574-5p expression has been shown to increase C18-ceramide levels in multiple human cancer cell lines [41].

Serine-palmitoyl transferase (SPT), which converts L-serine and palmitoyl-CoA to 3-ketosphinganine, is the first and rate-limiting enzyme of the de novo biosynthetic pathway of ceramide and SM. SPT has been implicated in the pathogenesis of atherosclerosis and its modulation has been suggested to be a novel therapeutic target in atherosclerosis [42]. Several studies have shown that the pharmacological targeting of SPT through the inhibition with myriocin protects from atherosclerosis in ApoE knockout mice [42–45]. Myriocin has been shown to reduce the levels of not only ceramide but other sphingolipids downstream of ceramide, such as SM, glucosylceramide and S1P as well [42,44,45]. Similar to pharmacological intervention, targeting the enzymes involved in sphingolipid biosynthesis through miRNAs may be a potential therapeutic intervention for alleviating atherosclerosis. Serine-palmitoyl transferase long chain base subunit 1 (SPTLC1) and 2 (SPTLC2) are the subunits of SPT and one study found a negative correlation between the expression levels of miR-137/-181c and SPTLC1 [40]. Transfection of primary rat astrocytes with miR-137 and miR-181c showed a significant suppression of the endogenous SPTLC1 expression and cellular ceramide levels. On the other hand, anti-miR-137 and anti-miR-181c significantly increased the endogenous SPTLC1 expression and cellular ceramide levels. In the same study, a negative correlation was also found between miR-9/-29a/b-1 and SPTLC2 protein expression in sporadic Alzheimer's disease brains. Again, the transient transfection of primary rat astrocytes with miR-9, miR-29a, and miR-29b-1 significantly suppressed the endogenous SPTLC2 and cellular ceramide levels and their antagomirs significantly enhanced the expression of SPTLC2 and cellular ceramide levels.

Finally, plasma S1P levels have been reported to be lower in CVD patients, suggesting its involvement in the pathogenesis of atherosclerosis [15]. Sphingosine kinase is involved in the conversion of sphingosine to S1P that have distinct intracellular and extracellular functions [46,47]. Exogenously expressed miR-101 has been shown to down-regulate SPK mRNA and protein expression in colorectal cancer cells [39]. The downregulation of SPK has been shown to result in increased ceramide levels in miR-101 expressed cells. On the other hand, the expression of SPK was enhanced by treating HT-29 human colorectal adenocarcinoma cells with the antagomiR-101 that resulted in lower levels of ceramide [39].

4. Conclusions

Significant advances have been made in studying the role of miRNAs in lipid metabolism. However, there is a paucity of information on their role in sphingolipid metabolism. Several sphingolipid synthesis genes have been targeted as potential therapeutics for various metabolic disorders, such as atherosclerosis. Sphingomyelin synthase, sphingomyelinase, and glucosylceramide synthase are some of the key enzymes that have been implicated in atherosclerosis [9,10,12,14] (Figure 1). It is likely that studying the panel of miRNAs that regulate sphingolipid metabolism could help us to understand these pathways better, and the modulation of these pathways may prove to be a potential therapeutic strategy. There are only a few studies that have reported the regulation of enzymes involved in sphingolipid biosynthetic pathways by certain miRNAs [39–41], and most of these studies are targeted towards ceramide synthesis in various cancer cell lines [39,41]. However, there is a lack of information on whether miRNAs regulate sphingolipid metabolism in vivo and thereby influence atherosclerosis. Based on the genetic deficiency and pharmacological inhibition studies [9,10,12,14,42], we predict that the regulation of some of the key enzymes in sphingolipid biosynthetic pathways by miRNAs may be useful and a potential therapeutic strategy to alleviate atherosclerosis. Therefore, more focused studies are needed to identify and understand the role of various miRNAs in regulating sphingolipid metabolism.

Author Contributions: Z.J. collected materials and edited the manuscript. A.B. and J.I. conceived and wrote the manuscript. All the authors read and approved the manuscript.

Conflicts of Interest: The authors declare no conflict of interest.

Abbreviations

ABCA1	ATP-binding cassette transporter A1
ACACA	acetyl-CoA carboxylase alpha
ACC2	acetyl-coA carboxylase 2
ACLY	ATP citrate lyase
ApoA1	apolipoprotein A1
CerS	ceramide synthase
CPT1A	carnitine palmitoyltransferase 1A
CROT	carnitine *O*-octanoyltransferase
CVD	cardiovascular disease
ELOVL5	ELOVL fatty acid elongase 5
FASN	fatty acid synthase
HADHB	hydroxyacyl-CoA dehydrogenase trifunctional multienzyme complex subunit beta
HDL	high-density lipoprotein
LDL	low-density lipoprotein
LPGAT1	lysophosphatidylglycerol acyltransferase 1
LXRα	liver X receptor alpha
MBOAT1	membrane bound *O*-acyltransferase domain containing 1
miRNAs	microRNAs
MTTP	microsomal triglyceride transfer protein
SCD1	steroyl-coA desaturase 1
SM	sphingomyelin
SMS	sphingomyelin synthase
S1P	sphingosine-1-phosphate
SPK	sphingosine kinase
SPT	serine-palmitoyl transferase
SPTLC1	serine-palmitoyl transferase long chain base subunit 1
SPTLC2	serine-palmitoyl transferase long chain base subunit 2
SR-B1	scavenger receptor class B type 1
nSMase2	type 2-neutral sphingomyelinase
SREBF1	sterol regulatory element binding factor 1
STARD3	stAR related lipid transfer domain containing 3
3′-UTR	3′ untranslated region
VLDL	very low-density lipoprotein

References

1. Glass, C.K.; Witztum, J.L. Atherosclerosis. The road ahead. *Cell* **2001**, *104*, 503–516. [CrossRef]
2. Lusis, A.J. Atherosclerosis. *Nature* **2000**, *407*, 233–241. [CrossRef] [PubMed]
3. Borodzicz, S.; Czarzasta, K.; Kuch, M.; Cudnoch-Jedrzejewska, A. Sphingolipids in cardiovascular diseases and metabolic disorders. *Lipids Health Dis.* **2015**, *14*, 55. [CrossRef] [PubMed]
4. Iqbal, J.; Walsh, M.T.; Hammad, S.M.; Hussain, M.M. Sphingolipids and Lipoproteins in Health and Metabolic Disorders. *Trends Endocrinol. Metab.* **2017**, *28*, 506–518. [CrossRef] [PubMed]
5. Pewzner-Jung, Y.; Ben-Dor, S.; Futerman, A.H. When do Lasses (longevity assurance genes) become CerS (ceramide synthases)? Insights into the regulation of ceramide synthesis. *J. Biol. Chem.* **2006**, *281*, 25001–25005. [CrossRef] [PubMed]

6. Ogretmen, B.; Hannun, Y.A. Biologically active sphingolipids in cancer pathogenesis and treatment. *Nat. Rev. Cancer* **2004**, *4*, 604–616. [CrossRef] [PubMed]

7. Iqbal, J.; Walsh, M.T.; Hammad, S.M.; Cuchel, M.; Tarugi, P.; Hegele, R.A.; Davidson, N.O.; Rader, D.J.; Klein, R.L.; Hussain, M.M. Microsomal triglyceride transfer protein transfers and determines plasma concentrations of ceramide and sphingomyelin but not glycosylceramide. *J. Biol. Chem.* **2015**, *290*, 25863–25875. [CrossRef] [PubMed]

8. Jiang, X.C.; Paultre, F.; Pearson, T.A.; Reed, R.G.; Francis, C.K.; Lin, M.; Berglund, L.; Tall, A.R. Plasma sphingomyelin level as a risk factor for coronary artery disease. *Arterioscler. Thromb. Vasc. Biol.* **2000**, *20*, 2614–2618. [CrossRef] [PubMed]

9. Li, Z.; Fan, Y.; Liu, J.; Li, Y.; Huan, C.; Bui, H.H.; Kuo, M.S.; Park, T.S.; Cao, G.; Jiang, X.C. Impact of sphingomyelin synthase 1 deficiency on sphingolipid metabolism and atherosclerosis in mice. *Arterioscler. Thromb. Vasc. Biol.* **2012**, *32*, 1577–1584. [CrossRef] [PubMed]

10. Liu, J.; Huan, C.; Chakraborty, M.; Zhang, H.; Lu, D.; Kuo, M.S.; Cao, G.; Jiang, X.C. Macrophage sphingomyelin synthase 2 deficiency decreases atherosclerosis in mice. *Circ. Res.* **2009**, *105*, 295–303. [CrossRef] [PubMed]

11. Kasumov, T.; Li, L.; Li, M.; Gulshan, K.; Kirwan, J.P.; Liu, X.; Previs, S.; Willard, B.; Smith, J.D.; McCullough, A. Ceramide as a mediator of non-alcoholic Fatty liver disease and associated atherosclerosis. *PLoS. ONE* **2015**, *10*, e0126910. [CrossRef] [PubMed]

12. Lallemand, T.; Rouahi, M.; Swiader, A.; Grazide, M.H.; Geoffre, N.; Alayrac, P.; Recazens, E.; Coste, A.; Salvayre, R.; Negre-Salvayre, A.; et al. nSMase2 (Type 2-Neutral Sphingomyelinase) Deficiency or Inhibition by GW4869 Reduces Inflammation and Atherosclerosis in Apoe $(-/-)$ Mice. *Arterioscler. Thromb. Vasc. Biol.* **2018**, *38*, 1479–1492. [CrossRef] [PubMed]

13. Dawson, G.; Kruski, A.W.; Scanu, A.M. Distribution of glycosphingolipids in the serum lipoproteins of normal human subjects and patients with hypo- and hyperlipidemias. *J. Lipid Res.* **1976**, *17*, 125–131. [PubMed]

14. Chatterjee, S.; Bedja, D.; Mishra, S.; Amuzie, C.; Avolio, A.; Kass, D.A.; Berkowitz, D.; Renehan, M. Inhibition of glycosphingolipid synthesis ameliorates atherosclerosis and arterial stiffness in apolipoprotein E-/- mice and rabbits fed a high-fat and -cholesterol diet. *Circulation* **2014**, *129*, 2403–2413. [CrossRef] [PubMed]

15. Knapp, M.; Lisowska, A.; Zabielski, P.; Musial, W.; Baranowski, M. Sustained decrease in plasma sphingosine-1-phosphate concentration and its accumulation in blood cells in acute myocardial infarction. *Prostag. Other Lipid Mediat.* **2013**, *106*, 53–61. [CrossRef] [PubMed]

16. Egom, E.E.; Mamas, M.A.; Chacko, S.; Stringer, S.E.; Charlton-Menys, V.; El-Omar, M.; Chirico, D.; Clarke, B.; Neyses, L.; Cruickshank, J.K.; et al. Serum sphingolipids level as a novel potential marker for early detection of human myocardial ischaemic injury. *Front. Physiol.* **2013**, *4*, 130. [CrossRef] [PubMed]

17. Sattler, K.; Lehmann, I.; Graler, M.; Brocker-Preuss, M.; Erbel, R.; Heusch, G.; Levkau, B. HDL-bound sphingosine 1-phosphate (S1P) predicts the severity of coronary artery atherosclerosis. *Cell. Physiol. Biochem.* **2014**, *34*, 172–184. [CrossRef] [PubMed]

18. Kurano, M.; Yatomi, Y. Sphingosine 1-Phosphate and Atherosclerosis. *J. Atheroscler. Thromb.* **2018**, *25*, 16–26. [CrossRef] [PubMed]

19. Santulli, G. Effects of low-carbohydrate and low-fat diets. *Ann. Intern. Med.* **2015**, *162*, 392. [CrossRef] [PubMed]

20. Christian, P.; Su, Q. MicroRNA regulation of mitochondrial and ER stress signaling pathways: Implications for lipoprotein metabolism in metabolic syndrome. *Am. J. Physiol. Endocrinol. Metab.* **2014**, *307*, E729–E737. [CrossRef] [PubMed]

21. Novak, J.; Bienertova-Vasku, J.; Kara, T.; Novak, M. MicroRNAs involved in the lipid metabolism and their possible implications for atherosclerosis development and treatment. *Mediat. Inflamm.* **2014**, *2014*, 275867. [CrossRef] [PubMed]

22. Wronska, A.; Kurkowska-Jastrzebska, I.; Santulli, G. Application of microRNAs in diagnosis and treatment of cardiovascular disease. *Acta Physiol. (Oxf.)* **2015**, *213*, 60–83. [CrossRef] [PubMed]

23. Bartel, D.P. MicroRNAs: Target recognition and regulatory functions. *Cell* **2009**, *136*, 215–233. [CrossRef] [PubMed]

24. Filipowicz, W.; Bhattacharyya, S.N.; Sonenberg, N. Mechanisms of post-transcriptional regulation by microRNAs: Are the answers in sight? *Nat. Rev. Genet.* **2008**, *9*, 102–114. [CrossRef] [PubMed]

25. Wilczynska, A.; Bushell, M. The complexity of miRNA-mediated repression. *Cell Death Differ.* **2015**, *22*, 22–33. [CrossRef] [PubMed]

26. Lim, L.P.; Lau, N.C.; Garrett-Engele, P.; Grimson, A.; Schelter, J.M.; Castle, J.; Bartel, D.P.; Linsley, P.S.; Johnson, J.M. Microarray analysis shows that some microRNAs downregulate large numbers of target mRNAs. *Nature* **2005**, *433*, 769–773. [CrossRef] [PubMed]

27. Ha, T.Y. MicroRNAs in Human Diseases: From Cancer to Cardiovascular Disease. *Immune Netw.* **2011**, *11*, 135–154. [CrossRef] [PubMed]

28. Scholer, N.; Langer, C.; Dohner, H.; Buske, C.; Kuchenbauer, F. Serum microRNAs as a novel class of biomarkers: A comprehensive review of the literature. *Exp. Hematol.* **2010**, *38*, 1126–1130. [CrossRef] [PubMed]

29. Olson, E.N. MicroRNAs as therapeutic targets and biomarkers of cardiovascular disease. *Sci. Transl. Med.* **2014**, *6*, 239ps3. [CrossRef] [PubMed]

30. Aryal, B.; Singh, A.K.; Rotllan, N.; Price, N.; Fernandez-Hernando, C. MicroRNAs and lipid metabolism. *Curr. Opin. Lipidol.* **2017**, *28*, 273–280. [CrossRef] [PubMed]

31. Zaiou, M.; Bakillah, A. Epigenetic Regulation of ATP-Binding Cassette Protein A1 (ABCA1) Gene Expression: A New Era to Alleviate Atherosclerotic Cardiovascular Disease. *Diseases* **2018**, *6*, 34. [CrossRef] [PubMed]

32. Rayner, K.J.; Suarez, Y.; Davalos, A.; Parathath, S.; Fitzgerald, M.L.; Tamehiro, N.; Fisher, E.A.; Moore, K.J.; Fernandez-Hernando, C. MiR-33 contributes to the regulation of cholesterol homeostasis. *Science* **2010**, *328*, 1570–1573. [CrossRef] [PubMed]

33. Sud, N.; Taher, J.; Su, Q. MicroRNAs and Noncoding RNAs in Hepatic Lipid and Lipoprotein Metabolism: Potential Therapeutic Targets of Metabolic Disorders. *Drug Dev. Res.* **2015**, *76*, 318–327. [CrossRef] [PubMed]

34. Novak, J.; Olejnickova, V.; Tkacova, N.; Santulli, G. Mechanistic Role of MicroRNAs in Coupling Lipid Metabolism and Atherosclerosis. *Adv. Exp. Med. Biol.* **2015**, *887*, 79–100. [PubMed]

35. Soh, J.; Iqbal, J.; Queiroz, J.; Fernandez-Hernando, C.; Hussain, M.M. MicroRNA-30c reduces hyperlipidemia and atherosclerosis in mice by decreasing lipid synthesis and lipoprotein secretion. *Nat. Med.* **2013**, *19*, 892–900. [CrossRef] [PubMed]

36. Esau, C.; Davis, S.; Murray, S.F.; Yu, X.X.; Pandey, S.K.; Pear, M.; Watts, L.; Booten, S.L.; Graham, M.; McKay, R.; et al. miR-122 regulation of lipid metabolism revealed by in vivo antisense targeting. *Cell Metab.* **2006**, *3*, 87–98. [CrossRef] [PubMed]

37. Rayner, K.J.; Esau, C.C.; Hussain, F.N.; McDaniel, A.L.; Marshall, S.M.; van Gils, J.M.; Ray, T.D.; Sheedy, F.J.; Goedeke, L.; Liu, X.; et al. Inhibition of miR-33a/b in non-human primates raises plasma HDL and lowers VLDL triglycerides. *Nature* **2011**, *478*, 404–407. [CrossRef] [PubMed]

38. Miller, A.M.; Gilchrist, D.S.; Nijjar, J.; Araldi, E.; Ramirez, C.M.; Lavery, C.A.; Fernandez-Hernando, C.; McInnes, I.B.; Kurowska-Stolarska, M. MiR-155 has a protective role in the development of non-alcoholic hepatosteatosis in mice. *PLoS ONE* **2013**, *8*, e72324. [CrossRef] [PubMed]

39. Chen, M.B.; Yang, L.; Lu, P.H.; Fu, X.L.; Zhang, Y.; Zhu, Y.Q.; Tian, Y. MicroRNA-101 down-regulates sphingosine kinase 1 in colorectal cancer cells. *Biochem. Biophys. Res. Commun.* **2015**, *463*, 954–960. [CrossRef] [PubMed]

40. Geekiyanage, H.; Chan, C. MicroRNA-137/181c regulates serine palmitoyltransferase and in turn amyloid beta, novel targets in sporadic Alzheimer's disease. *J. Neurosci.* **2011**, *31*, 14820–14830. [CrossRef] [PubMed]

41. Meyers-Needham, M.; Ponnusamy, S.; Gencer, S.; Jiang, W.; Thomas, R.J.; Senkal, C.E.; Ogretmen, B. Concerted functions of HDAC1 and microRNA-574-5p repress alternatively spliced ceramide synthase 1 expression in human cancer cells. *EMBO Mol. Med.* **2012**, *4*, 78–92. [CrossRef] [PubMed]

42. Hojjati, M.R.; Li, Z.; Zhou, H.; Tang, S.; Huan, C.; Ooi, E.; Lu, S.; Jiang, X.C. Effect of myriocin on plasma sphingolipid metabolism and atherosclerosis in apoE-deficient mice. *J. Biol. Chem.* **2005**, *280*, 10284–10289. [CrossRef] [PubMed]

43. Park, T.S.; Panek, R.L.; Mueller, S.B.; Hanselman, J.C.; Rosebury, W.S.; Robertson, A.W.; Kindt, E.K.; Homan, R.; Karathanasis, S.K.; Rekhter, M.D. Inhibition of sphingomyelin synthesis reduces atherogenesis in apolipoprotein E-knockout mice. *Circulation* **2004**, *110*, 3465–3471. [CrossRef] [PubMed]

44. Glaros, E.N.; Kim, W.S.; Wu, B.J.; Suarna, C.; Quinn, C.M.; Rye, K.A.; Stocker, R.; Jessup, W.; Garner, B. Inhibition of atherosclerosis by the serine palmitoyl transferase inhibitor myriocin is associated with reduced plasma glycosphingolipid concentration. *Biochem. Pharmacol.* **2007**, *73*, 1340–1346. [CrossRef] [PubMed]
45. Glaros, E.N.; Kim, W.S.; Quinn, C.M.; Jessup, W.; Rye, K.A.; Garner, B. Myriocin slows the progression of established atherosclerotic lesions in apolipoprotein E gene knockout mice. *J. Lipid Res.* **2008**, *49*, 324–331. [CrossRef] [PubMed]
46. Shida, D.; Takabe, K.; Kapitonov, D.; Milstien, S.; Spiegel, S. Targeting SphK1 as a new strategy against cancer. *Curr. Drug Targets* **2008**, *9*, 662–673. [CrossRef] [PubMed]
47. Vadas, M.; Xia, P.; McCaughan, G.; Gamble, J. The role of sphingosine kinase 1 in cancer: Oncogene or non-oncogene addiction? *Biochim. Biophys. Acta* **2008**, *1781*, 442–447. [CrossRef] [PubMed]

MDPI

St. Alban-Anlage 66

4052 Basel

Switzerland

Tel. +41 61 683 77 34

Fax +41 61 302 89 18

www.mdpi.com

Diseases Editorial Office

E-mail: diseases@mdpi.com

www.mdpi.com/journal/diseases